Scribe Publications
OUR AGEING BRAIN

ANDRÉ ALEMAN is professor of cognitive neuro-psychology at the University of Groningen. An internationally respected neuroscientist, he has received international awards and scholarships for his work. In 2011, he published the highly successful *Figments of Our Imagination: why we see, hear, and think things that aren't there* (published in Dutch).

ANNETTE MILLS was born in the UK. After graduating from King's College, London University, she lived for several years in North Africa and the Middle East, finally settling in the Netherlands in 1978. She has worked as a translator since 1981, both at the Dutch Ministry of Foreign Affairs and as a freelancer.

OUR AGEING BRAIN

how our mental capacities develop
as we grow older

André Aleman

Translated by Annette Mills

SCRIBE
Melbourne • London

Scribe Publications
18–20 Edward St, Brunswick, Victoria 3056, Australia
2 John St, Clerkenwell, London, WC1N 2ES, United Kingdom

First published in Dutch as *Het Seniorenbrein* by Uitgeverij Atlas Contact, Amsterdam, 2012

Published by Scribe 2014
Reprinted 2015

The publisher gratefully acknowledges the support of the Dutch Foundation for Literature.

Nederlands letterenfonds dutch foundation for literature

Typeset in Minion 11.5/14 pt by the publishers
Printed and bound in Australia by Griffin Press

 The paper this book is printed on is certified against the Forest Stewardship Council® Standards. Griffin Press holds FSC chain of custody certification SGS-COC-005088. FSC promotes environmentally responsible, socially beneficial and economically viable management of the world's forests.

National Library of Australia Cataloguing-in-Publication data

Aleman, André, author.

Our Ageing Brain: how our mental capacities develop as we grow older / André Aleman; Annette Mills (translator).

9781925106114 (Australian edition)
9781922247636 (UK edition)
9781925113259 (e-book)

1. Age and intelligence. 2. Ability, Influence of age on. 3. Human information processing–Age factors. 4. Cognition in old age. 5. Aging–Psychological aspects.

Other Authors/Contributors: Mills, Annette, translator.

155.67

scribepublications.com.au
scribepublications.co.uk

CONTENTS

Introduction

In 2012, at the age of 102, Theodora Claassen-Roos told the newspaper *NRC Handelsblad* that for the last few years, the mayor had paid her a visit on her birthday. But she had told her children (the oldest is 77; the youngest, 64) that she really didn't need to celebrate her birthday anymore. 'Far too expensive to do that every year,' she said. 'It's just a waste of money.'

'Theodora has always led a healthy life. 'I go for a little walk every day. It keeps me active. I always meet people I know and we have a chat,' she explained. 'But apart from that, I don't know what I've done to live so long.' She reads two newspapers each day, and has a particular interest in the articles about science. 'But I don't spend the whole day reading, you know. That would be a real waste of time!'

Why is one person still fit and healthy at the age of 100, while another might have serious memory problems in their sixties? Is everyone at risk of Alzheimer's? What does the brain of an 80-year-old look like? And are there any advantages to having an older brain? In this book, I try to answer these questions on the basis of the latest scientific findings.

Since university, I have been fascinated by ageing processes in the brain. My final paper at the University Medical Centre Utrecht was on the link between growth hormone and cognitive skills in older men, and my article on the subject was published in *The Journal of Clinical Endocrinology and Metabolism*. In the years that followed, I was involved in a range of studies of brain function in ageing subjects. Now, in *Our Ageing Brain*, I report on these studies and on all kinds of research performed by colleagues in this field. My current employer, the University Medical Centre Groningen, focuses on research into healthy ageing. There, in 2012, I started on a new study into brain function in older people who suffer from forgetfulness but have not been diagnosed with dementia, since they are still capable of leading an independent life.

In this book, I often refer to 'older people' or 'seniors'. By that, I mean people over the age of 65, a dividing line frequently adopted in medical research. It is sometimes called 'the third age', following on from the first age (youth to early twenties) and the second age (middle age). So I'm using what has long been seen as the retirement age as a marker. In fact, there's also a solid argument for regarding 70 as the beginning of 'old age', since people currently live longer than they did 50 years ago and stay fit for longer. But this book is not just for those over 65. I have written it for everyone who wants to know what research has actually established about the brain and ageing.

In 2012, there were 3.22 million people over 65 in Australia, comprising 14 per cent of the total population. According to the Australian Bureau of Statistics, this figure is predicted to rise to between 23 and 25 per cent of the population by 2056. Since 2010, the group known as baby boomers (people born in the post-war 'baby boom' between 1946 and 1964) have started to reach retirement age. And they are living longer: the life expectancy of Australians continues to grow. In 2012, the average 65-year-old woman could expect to live to the age of 87 (or 1.2 years more than in 2002); and the average 65-year-old man, to the age of 82 (or 1.7 years more than in 2002). Thus, the population is 'greying' rapidly. And the central focus of this book is what happens to our 'grey cells' in that process.

As we age, our brain cells undergo an irreversible decline. Some brain cells shrink, connections between different areas of the brain disappear, memory and concentration erode, and other cognitive abilities slow down. But it's not all bad news. Older people are often happier than their younger counterparts. They cope better with emotions and stress, and are better at making complex decisions. There are, of course, huge differences between people in this age group, and I try to explain why this is so. On the basis of the latest research findings, I'll show you what changes occur in the brain and how older people use other areas of the brain to compensate for decline. I'll also take a look at the enormous variety of pills, powders, and supplements that promise to reverse the process but actually deliver very little. And I'll tell you what you can do that does help. Finally, I'll deal with the intriguing question of how we become wiser thanks to decline, and what 'successful ageing' — a term that has even made it into the scientific literature — really is.

I would like to thank Ine Soepnel, my editor at Atlas Contact, for her advice and requests for clarification, which made this a much better book. My thanks to Anita Roeland for her contribution to the process, and to Berber Munstra, who did the illustrations. And finally, a special thank-you to Finnie, my wife, for her support. I'm looking forward to growing old and grey with her. Bring on that ageing brain!

1

'Everything Goes So Fast These Days':

how our mental capacities change

After turning 50, almost all of us worry from time to time that our memories are beginning to fail. Perhaps you can't recall someone's name, or where you left the house keys. And that's just the beginning — before you know it, you're forgetting to turn off the gas … Or you might find it difficult to keep up with technological progress; today's fast-paced information society is leaving you behind. Twitter, Facebook, Google+, iPhone, iPad, BlackBerry: none of these concepts existed 15 years ago. Now they are common currency.

Everyone wants to live for a long time, but no one wants to be old. If you ask 40-year-olds if they would rather be 65, almost none of them will say yes, even though they see the advantages of more leisure time after retirement. Besides poorer health, the greatest fear is of memory loss and a declining ability to concentrate. Or people worry that they won't be able to keep up as their thought processes become slower and less nimble. As long ago as the 4th century BCE, the Greek philosopher Plato maintained that the loss of physical strength caused by ageing was accompanied by a reduction in intellectual capacity. At an advanced age, he argued, you are no more capable of learning new things than of running as fast as you once did. Was Plato right? Which mental capacities do in fact deteriorate with age? When does the process start? How fast does it progress? And what capacities remain unaffected?

PRECONCEPTIONS

Many people, including older people, have an inaccurate and far too bleak a picture of life after 65, or the 'third age' as it is known. In 2008, the theme of the Dutch *Boekenweek* (Literature Week) was 'On the Elderly: the third age and literature'. To mark the occasion, the newspaper *de Volkskrant* and the VU University Amsterdam conducted a survey on perceptions of older people in the Netherlands. Participants had to say whether a statement was correct or incorrect. One example was 'Fifty per cent of older people are lonely'. Of the 751 participants, 47 per cent thought this statement was correct, but in fact they were wrong. Only a small percentage of older people are lonely, as shown in the long-term, large-scale Longitudinal Ageing Study Amsterdam, which supplied the information for the survey. The participants got many of the other answers wrong too, their view being consistently gloomier than reality. For instance, they often made the wrong choice in the statements regarding declining networks of family, friends, and acquaintances (13 per cent got this right). Most of the participants thought these networks shrink drastically as a person ages. In reality, there are many differences between individuals, and it is only at a very advanced age that the average size of people's networks begins to decline. Most of the participants (75 per cent) also believed that over the last 15 years, visits from children had declined, and that in the last ten years, older people had developed a healthier lifestyle (58 per cent). Both statements are incorrect. Visits between older people and their children have increased rather than decreased, while the lifestyle of younger seniors is less healthy than that of their predecessors. The frequency of depression and conservative attitudes among older people was overestimated; the degree of sexual activity, underestimated. Statements regarding health were, on the whole, more correctly assessed than those regarding social functioning.

This negative view of older people is not without its dangers. It can become a self-fulfilling prophecy. Research has shown that a positive view of ageing has a favourable effect on healthy life expectancy. Indeed, this effect is stronger than the effect of lifestyle factors such as physical activity, smoking, or obesity. One study compared participants' mortality rates with the answers they had given in a survey conducted dozens of years previously. People with a more positive perception of ageing lived on average 7.5 years longer than those whose view was negative. The researchers concluded that a positive attitude reduced stress and encouraged involvement in activities that enhance wellbeing. Another study investigated whether a positive attitude influenced the risk of mortality in 1993 in a population of 6,856 people who had completed a questionnaire in 1965. Here, too, the mortality risk was lower for people with a positive attitude. Social networks played a major role: people with a positive attitude had more social contact and lived longer. Finally, American research showed that positive expectations about growing older in people aged around 60 predicted greater happiness at the ages of 70 and 80. These results remained valid when controlled for differences between individuals' natural dispositions (some people are 'happier' by nature than others) before the age of 60, and for income and health (which of course have an effect on your happiness when you're 70). We'll take a closer look at how optimism helps the brain function better — particularly in dealing with stress and setbacks — in a later chapter. In fact, reading this book may prolong your life simply because it focuses your attention on the positive aspects of growing older.

Negative stereotypes, on the other hand, lead to lower performance and a decline in wellbeing. If you ask a group of older people to read a number of words related to ageing that have negative connotations (for instance, 'senile'), and ask them to take a memory test immediately afterwards, they perform less well than if they have first read a list of words with positive connotations (such as 'wise').

The typical features of ageing may also exert a — probably subconscious — influence on us because they are associated with decline. In order to test this theory, Harvard professor Ellen Langer studied the effects of numerous signs of ageing. One of her predictions was that men who grew bald prematurely would suffer earlier from age-related disorders. Baldness is associated with ageing, and people confronted on a daily basis with baldness might feel older than they are. Her research did in fact establish a correlation between premature baldness and early onset of age-related health problems.

Another expected correlation was between clothing and feeling older: people who are 60 dress differently from those who are 25. In occupations where people wear uniforms (train conductors, police officers), this should not apply: everyone wears the same uniform, so there is no way of discerning one's age from one's clothing. A 20-year-old in uniform may feel older than he or she is, and a 65-year-old may feel younger. The study revealed that older uniformed employees have fewer age-related health problems than those who do not wear uniforms.

Langer also looked at spousal age differences. She wondered whether a person with a partner who was at least ten years younger had a more youthful lifestyle and whether they aged less quickly. This proved to be the case. On the other hand, a younger partner aged more quickly than someone married to a person of their own age.

Before conducting this study on age cues, Professor Langer had researched cultural influences on memory loss. She discovered that older people in rural China suffer less from memory problems than we do. Langer put this down to their lack of exposure to the negative stereotyping around age that is common in Western cultures. Her research showed that deaf Americans (who are also less exposed to negative stereotyping around age) exhibited fewer signs of failing memory. In short, the individual's response to growing older and the influence exerted by widely accepted ideas both make a difference. A positive attitude — often perfectly justified, since many older people are in good health — keeps you young.

Aldo Ciccolini is an accomplished concert pianist. He plays in prestigious concert halls and receives enthusiastic reviews. At a concert in Rotterdam on 9 May 2011, he played works by Mozart and, after the interval, works by Liszt. As an encore, he played several pieces, including *Spanish Dance No. 5* by Enrique Granados. What made this concert so special that it is worth mentioning here? The answer is that Ciccolini was 85 years old when he gave it. The music critic in the Dutch newspaper *Trouw* was full of praise.[1] His review was headed: 'Eighty-five-year-old Ciccolini's Liszt sublime and unforgettable'. He went on: 'In his melodic, perfectly elaborated interpretation, he developed enormous power where needed'. According to the critic, the encores showed that Ciccolini was still a virtuoso. Did his advanced age have no influence at all on his performance? Perhaps a little. The reviewer described how the pianist was compelled to 'rescue himself from a failure of memory' in Mozart's *Sonata in B-flat Major* (KV 333), but he did it 'so amazingly well that few people listening would have noticed'. So it appears that his memory did fail him at one point, for reasons undoubtedly connected with his age. Yet he still managed to correct this in such a masterly way that virtually no one noticed.

This story shows that even the memories of people who still perform exceptionally well at an advanced age are beginning to show signs of wear and tear. When we think about the deterioration in our mental faculties as the years go by, memory is the first thing that comes to mind. It has long been known that older people have more difficulty in storing new information than their younger counterparts. The Greek philosopher Aristotle employed the metaphor of a wax tablet. A new tablet is soft and easy to write on. The wax hardens as it ages, making it increasingly difficult to carve. In the same way, it's harder for older people to retain new impressions. An ichthyologist (a zoologist specialising in the study of fish) who became the dean of a university department claimed that for every name of a new student he had to store in his memory, the name of a fish species disappeared.[2] The idea behind this is that the memory is a finite space: once it is full, something has to be removed before you can add new items. However logical this may seem, a great deal of research has been done into the capacity of our long-term memory, and there is no proof that it can become 'full'. On the contrary, it's amazing just how much information can be held by — and even added to — our long-term memory as it ages.

Whether this information is stored and whether you can access it easily are two different issues. Some information just never reaches our memory banks, so you can never retrieve it. Take the example of someone who introduces himself at a crowded reception: 'Hello, I'm John Smith.' As he does so, you're distracted by someone bumping into you and by the mere fact of introducing yourself. Because you weren't really paying attention, his name isn't stored in your memory. It's hardly surprising if some weeks later you can't recall it. In fact, you never actually knew his name. To remember a person's name, you have to pay attention and, preferably, do something with that information. For example, you can link his name with that of someone you know, and create an image in your mind of both Johns side by side. The chances are you will retain his name — or at least his first name!

Many people associate ageing with forgetfulness. The general impression is that forgetfulness begins around the age of 60, or a couple of years earlier. But is this true? Most of my acquaintances who are about 60 don't appear to have any serious memory problems. These are more likely in people aged 75 and over. Yet research has established a very different starting date for memory decline. And that — prepare yourself — is around your twentieth year. Between the ages of 60 and 70, your average memory performance declines more rapidly (see figure 1).

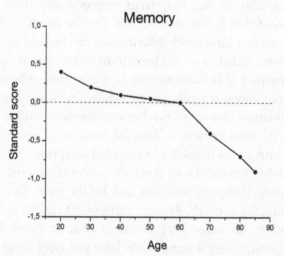

FIGURE 1. *Memory performance gradually declines with age.*

This decline is partly attributable to the fact that people of this age often make less use of their mental abilities (after retirement, for instance). But most of them are not suffering from a memory disorder — their score differs by less than one point from the population average (the baseline).

Incidentally, the word 'memory' is a misleading one. It gives the impression that we are talking about a single entity or a single mental skill. In fact, research over the last 50 years has established that there are different memory processes. For example, we all have short-term and long-term memory. As everyone knows, your short-term memory is for things you need to remember for less than a minute, such as a telephone number you have to call. In that process, it's important not to start thinking about something else or you'll forget the number. This applies to both young and old, though slightly more to older people. Long-term memory is responsible for everything you need to retain for over a minute, even if other things have occupied your attention in between. Then there's working memory, involved in processes taking place in your short-term memory — remembering, for example, a number that changes, as when you're adding and subtracting figures.

Long-term memory is divided into procedural memory and declarative memory. Procedural memory relates to actions such as cycling and playing piano. You learned at some stage how to do it, and the necessary body movements are directed by your procedural memory. These unconscious processes, which through practice have become automatic, are vital. If you've done a lot of cycling, you no longer have to think consciously about how you move forward or take a corner.

Declarative memory, on the other hand, involves the conscious recall of information, such as a shopping list you've memorised. This form of memory can be either verbal or visual, and is further subdivided into semantic memory and episodic memory. The first relates to the meaning of concepts (and people's names); the second, to events. To illustrate: knowing when you last went for a bike ride draws on your episodic memory, knowing what a bike is involves your semantic memory, and knowing how to ride a bike depends on your procedural memory. A subset of episodic memory is autobiographical memory, which relates to life events and experiences (see figure 2).

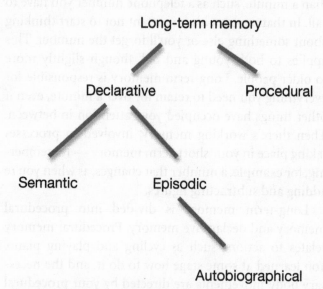

FIGURE 2. *The subsets of long-term memory.*

Finally, we come to prospective memory, which retains the things you still have to do, such as ringing the garage, or picking up flowers before you go and visit your aunt. Or cleaning out the cat litter tray. To discover which memory functions decline with age and which do not, let's look more closely at the major memory processes.

'What was the name of Peter's brother-in-law? You know, we met him at that party ...' Everyone occasionally forgets a name. As we age, this problem can recur more frequently. It's as though the information is hidden at some deep level, in a drawer of an old cupboard. The drawer is stuck, or you don't have the key. The information was stored in your memory, but you can't access it anymore. Ageing affects episodic and working memory the most. Psychologists often use a word list to test episodic memory. They read aloud 15 words that have nothing to do with one another: 'plant, beard, hat ...' and so on. The subject is then asked to repeat the list. Most middle-aged people (around 45 years old) can recall about seven words. The list is then read again four times, after which most people in this age group can remember 12 of the 15. After a quarter of an hour, the subject is asked to repeat as many as possible of the words in the list. This is more difficult. Most middle-aged people can recall no more than ten words. In people aged around 70, this drops to five (immediately after the first reading), nine (having heard the list five times) and seven (after 15 minutes). The ability to remember random words thus declines considerably as we age. Remembering a shopping list becomes more difficult, particularly if the items are unrelated. If you need six items for a curry that you plan to make, they're easier to remember than six unrelated items such as icing sugar, cherry yoghurt, toilet paper, leeks, macaroni, and oranges.

But what about remembering stories? This is also a task for your episodic memory. Stories, of course, have more structure and meaning than word lists, but does that make them easier for older people to retain? The Wechsler Memory Scale, which is used by psychologists all over the world, is a good measure of memory function based on storytelling. One of the components in the test involves listening to a story that the subjects immediately have to retell to the best of their abilities. It turns out that people aged around 70 are as capable of reproducing the story as people aged around 20. Thirty minutes after they have heard the story, they undergo another memory test, this time unannounced, in which they again have to reproduce as much as they can of the story. At this point, younger people can remember many more details than their seniors.

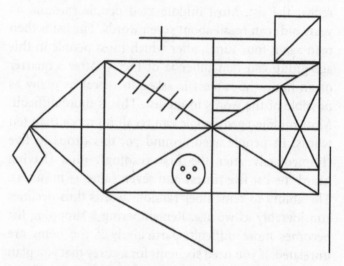

FIGURE 3. *Sample visual memory test. A complex diagram is copied, then drawn from memory 20 minutes later.*

Words and stories are verbal material stored in a person's verbal memory. Images are stored in our visual memory, which incidentally is not the same as a photographic memory (the ability to recall images almost perfectly). Visual memory is more general in nature: you can, for example, visualise your friend's house, but not the details. Does visual memory decline as quickly as verbal memory as people grow older? The answer is yes: the ability to recall recent visual information declines by about 30 per cent between the ages of 20 and 70. A common test for measuring this decline involves copying a complex diagram (see figure 3). Once the subject has copied the diagram, they have to draw it from memory. Twenty minutes later, they have to draw it again from memory. Once again, 20-year-olds can reproduce many more details than 70-year-olds.

Events are often stored in the form of a narrative. Many of the problems that occur as we age are verbal in nature: we forget names and stories. Yet there is also a lot of information that is visual. A useful way of remembering where you left your keys is to recall the image of you putting them down somewhere. The same applies to finding your way around. Even though we now have navigation systems in the car, there are still all kinds of situations where you have to use your visual memory — finding a department in a hospital that you've only visited once before, for instance, or a friend's flat in a large complex.

WORKING MEMORY

Your working memory retains the information that you need to perform a certain task. Cooking, for example, draws heavily on working memory. Like many other routine tasks, it involves having information available in your memory for a short while and, crucially, updating it as you go along. Imagine you're making a three-course meal. You have to prepare the vegetables, fry the meat, and get the oven ready for a casserole. Each of these actions requires certain steps to be carried out in a particular order, so you have to know precisely when the next step needs to be performed. This is quite a challenge for your working memory. The same applies to mental arithmetic or learning how to perform new, complex tasks, such as playing a musical instrument or using a computer program.

Research into working memory and ageing is generally based on tasks that resemble mental arithmetic. A common method is the n-back continuous performance task, in which the subject is presented with a sequence of numbers, one at a time, and has to indicate when a number corresponds to one that occurred n steps earlier in the sequence. If the subject has to remember a number that is one step back (that is, $n = 1$) they will be told, for example, to press a button when the number that is one step back is a 3. In the series 6, 3, 4, 5, ... they should have pressed the button when the 4 appeared, since the number one step before that was a 3. The load factor can be increased to make it more difficult. The subject has to remember a number two steps back ($n = 2$) or three ($n = 3$), and so on. For example, the instruction may be 'Press the button when the number two steps back was a 3'. In the sequence 7, 4, 3, 8, 1, 2 ... the number to be pressed is 1, since the number two steps back was a 3. Because the numbers appear one by one on the screen, the task is quite challenging, especially at the 3-back and 4-back levels. The number sequence in your working memory has to be constantly updated as new numbers appear, while at the same time you have to remember what the number was three steps back.

Most older people perform this task less well than their younger counterparts, particularly at the 3-back and 4-back levels. Various studies have shown that the main problem for older people lies in 'getting rid' of irrelevant information. As soon as a number is no longer relevant you have to ignore it, but this is difficult as it *was* relevant a short while before.

What this means is that besides being able to process fewer items simultaneously (retaining words in their verbal memory, for instance), older people are less able to repress unimportant information. Both aspects have a negative effect on working memory. That has consequences in our daily lives, especially when learning new tasks involving a large amount of information, not all of which is immediately relevant. This might be learning to work with a new computer program, or concentrating on a book, while your grandchildren are being particularly noisy. Both are more difficult for older people because they find it harder to ignore the irrelevant information (noise made by children) and that information is then present in their consciousness. This is related to a decline in the functioning of the prefrontal cortex (which is located at the front of the brain, and helps us filter out unimportant information). We'll come back to this later.

A simpler way to measure working memory is the 'self-ordered pointing test'. Here, the subject is given a file containing several sheets, each showing ten abstract designs (see figure 4 for examples of abstract designs).

Subjects are asked to point at one design per sheet. They are free to choose any of the designs, but can only choose each item once. They therefore have to retain the chosen items in their working memory. Again, this is more difficult for older people. Though it would appear to have little to do with day-to-day memory problems, it involves an important skill: the ability to distinguish between something you have already seen and something that is new. If you see a new painting at a friend's house, it's only good manners to comment on it the first time you see it. But if on every subsequent visit you start talking about the 'new painting' as if you've never seen it before, people will start to worry about your memory.

FIGURE 4. *Examples of abstract designs used in the self-ordered pointing test. The subject may point only once at a particular design and has to remember which items they have already indicated.*

FLEXIBLE THINKING

According to researchers at the University of Lorraine in Nancy, France, spiders become less proficient in spinning webs as they age. They used a common European house spider (*Zygiella x-notata*) as a model, and concluded that as they grow older, spiders begin to make mistakes. The researchers looked closely at the form, regularity, and number of holes in webs made by spiders of different ages. Webs woven by older spiders were less regular and contained large holes between the threads. Spiders that were eight months old (their life span is 12 months) produced webs with gaping holes and eccentric patterns. The researchers suggested that the poorer performance was due to a declining central nervous system, but said further research was needed. They believed that the study could shed more light on the underlying mechanisms of age-related decline in humans.

Do people become less proficient in complex tasks as they age? It is true that memory is not the only faculty that declines with age. What are known as 'executive functions' decline too. These are the basic skills people need to perform a task: the ability to organise, plan, and initiate actions; and to remain on track, control impulses, regulate emotions, adapt, and recover. You could describe them as the cognitive processes that co-ordinate, control, and manage our behaviour. They let you know what behaviour is appropriate to a specific situation and enable you to restrain or repress what is inappropriate. An example: if you are in a hurry to get somewhere, you have to repress the urge to keep driving when the traffic lights turn orange. Otherwise, you may end up driving through a red light. Another executive function allows you to concentrate on the task at hand despite distractions. Driving a car exposes you to all kinds of distractions, from beautiful buildings to attention-grabbing advertising to a call on your mobile phone. You have to be able to ignore these stimuli and concentrate on driving. Older people may find that difficult.

Let's go back to cooking, which is dependent on executive functions. The Canadian researcher Fergus Craik designed a virtual cookery experiment on the computer to compare the ability of younger and older subjects to plan the tasks involved in making an American breakfast. They were given an onscreen list of five items, from toast to pancakes, each requiring a different preparation time. The participants started the cooking process by selecting an image on a touch screen. The task was to have everything ready at the same time and to set the table while the food was cooking. Setting the table involved moving objects such as plates and cutlery from the edge of the screen to the correct place on the 'table'. There was a more difficult variant, in which each food item was on a separate computer screen, making it necessary to switch constantly from one screen to another to keep an eye on cooking times. Besides testing the ability to plan and maintain an overview of the situation, this task makes demands on prospective memory: remembering what you still have to do. By deducting the time needed, ideally, to get everything ready from the time someone has actually taken, you can measure how well that person has kept everything under control. The bigger the gap between the two, the more difficult it was for the subject to co-ordinate the various cooking times. The ability to plan and look ahead was measured by deducting the time the preparation of the first dish stopped from the time the preparation of the last dish came to an end. In an ideal situation, there should be no difference, since the instructions were to have the whole meal ready at the same time. The bigger the difference between these times, the harder it was for the person to perform the task.

What were the results? Older participants (with an average age of 70) clearly experienced greater difficulties than younger participants (with an average age of 20) in co-ordinating preparation times. They were also worse at planning but only in the more difficult version of the task, in which each product was shown on a different screen. These results match those from a number of other studies showing that older people have more difficulty with planning, though the other studies used more abstract tasks.

One such abstract task is the trail-making test. Here, the aim is to draw a line as quickly as possible between letters and numbers spread around a page as follows: 1–A–2–B–3–C– and so on. This is then compared with the performance of the same task involving only numbers, so 1–2–3– and so on. The time taken to perform the first task compared with the time taken on the second indicates how long it takes a person to switch between categories, in this case letters and numbers. Try this with figure 5.

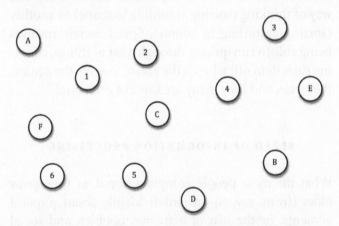

FIGURE 5. *Variant of the trail-making test. With a pencil, draw a line between the numbers and letters in the following order: 1–A–2–B– and so on. Complete the task as quickly as possible.*

Another test of mental or cognitive flexibility that you can do at home is a card game called SET. If you're playing this with your grandchildren, don't be surprised if you lose, despite the advantage you have in terms of general knowledge and probably years of experience in playing games. The simpler version of SET involves twelve cards with four features in three variations, laid out on a table. The features are colours, symbols, shadings, and the number of symbols. The players all try to make a set of three cards in which each individual feature is either the same on each card or different on each card.

Cognitive flexibility enables you to switch between categories (colour, shape, number) to decide whether there are enough cards of a certain type available. But SET doesn't only require flexibility. Speed is important, too. Which player is the first to see all the possible combinations? It could be that older people are just a little slower rather than less mentally flexible.

Flexibility involves being able to switch from one way of thinking (sorting according to shape) to another (sorting according to colour). Speed merely involves being able to run quickly through a list of things, checking each item off: where's the green, the red, the square, the circle, and how many are there of each one?

SPEED OF INFORMATION PROCESSING

What do most people complain about as they grow older (from, say, 70 onwards)? Mainly about physical ailments, or the loss of activities, hobbies, and social contact. The most frequent complaint in terms of mental capacity is a failing memory, followed by an inability to concentrate and being more easily distracted by external stimuli. Few people complain about how fast they can think. That is because speed of thought (referred to as 'speed of processing' by psychologists) is not an aim in itself but a way of achieving an aim — to understand the patient-information leaflet for your medication, for instance, or to remember the three items your neighbour asked you to bring from the shop. Yet many older people are aware that they need more time than younger people to process information. If it has to be processed quickly, they may have difficulty following it. But they have no problems at all if there is no time pressure. This raises the question of whether our mental faculties slow down as we age. We see this as perfectly normal where physical capacities are concerned: someone who is 75 can't walk as fast as someone who is 40. But does this apply to the brain?

The answer is yes, processing speed declines with age. It is, in fact, the most important faculty that declines. Timothy Salthouse, a prominent researcher in the field of cognitive ageing, suggests that slower thought processes are at the root of all other mental deterioration — for example, in memory and executive functions. This is a crucial point, since it goes to the heart of age-related decline.

How do you measure speed of information processing? There are two types of test: one that simply measures reaction time and one that focuses on how quickly a person performs a thought process. Both usually involve pressing a button or writing something down. Motor skills in the hand or finger also play a role in the first method. In a simple reaction-time test, the subject has to press one of five buttons as fast as possible when it lights up. That is straightforward. It gets more difficult if you have to press the button next to the one that lights up, since you have to repress the natural instinct to press the button that is lit up. The variant test usually takes more time, and people make more mistakes. A lot of research has been done into reaction time in the young and the old, and this shows that it declines slowly but surely from the age of 20. If this is represented on a graph, you see a straight line that begins to descend diagonally after that age.

Short reaction times are essential if you're driving, which is why older people need regular testing after the age of 70. This is best done in a relatively realistic setting, rather than by pressing buttons. Neuropsychologists at the University of Groningen use a driving simulator. The subject 'drives' the car and encounters realistic traffic situations that require fast braking or swerving. But there's more to driving than just braking in time for a pedestrian who suddenly crosses the road. Quick thinking is also essential for judging the complicated interplay of traffic, which, after all, is constantly changing. Imagine that you want to overtake on a busy highway. First, you have to quickly check whether there are vehicles behind you that are travelling faster than you are. Sometimes you have to decide in a split second whether you're going to pull out. But the fact that thinking slows with age doesn't necessarily mean that older people are by definition more dangerous drivers. That only applies to older people whose reaction time or executive function has been seriously eroded. Around 20 per cent of people over the age of 70 perform well below average when tested, but there are plenty of seniors who drive perfectly well. In March 2012, an 80-year-old South African woman received an award from the government because she had driven for 60 years without ever being fined for a traffic violation.

The transport minister said she was a role model for the South African population. Every year, reckless driving results in many deaths on South African roads.

Speed of processing (and reaction time, which is an aspect of this) is not only important when driving a car, but in many other situations where we are being bombarded with information — for example, if the plumber is explaining all the things your new central-heating system can do. The same applies in a stimuli-rich environment, such as a busy high street. Speed is also necessary to follow conversations involving a number of speakers; many people speak quickly and often start speaking before others have finished.

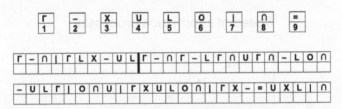

FIGURE 6. *Variant of the digit symbol substitution test: the corresponding digit has to be filled in under each symbol. See how far you get in 90 seconds.*

A common measure of how fast people think is the digit symbol substitution test (DSST), part of a range of intelligence tests (see figure 6). The test takes only a few minutes. Subjects are given a sheet of paper with a series of boxes and, at the top of the sheet, a code that links digits to abstract symbols. They are then instructed to use the code to write the correct digits in the blank boxes under a symbol. Say the code links the symbol Γ to the number 1, and the symbol | to the number 7, then you have to write that number under each occurrence of the symbol. It isn't difficult: the challenge lies in how far you can get, without making mistakes, in 90 seconds. A huge number of studies performed worldwide over the last 50 years have shown that this test can measure the earliest indications of age-related decline. It is highly sensitive, picking up small variations in cognitive skills. From the age of 20, performance deteriorates — subjects don't fill in as many correct digits — and this process continues (see figure 7). This is, in fact, the most reliable test of age-related cognitive decline, and it is used, together with other tests, by neuropsychologists all over the world in assessing older people who have been referred to them by a doctor.

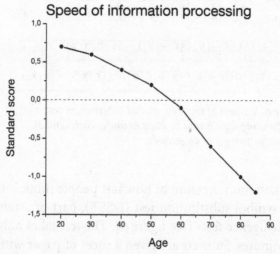

FIGURE 7. *Speed of information processing gradually declines as people age.*

But what does the DSST actually measure? You might expect that filling in the boxes requires more than just fast information processing. You have to write each number down, so it could also be a question of motor skills, which vary from person to person. And working memory comes into play: if you remember part of the code, you can work faster because you don't have to keep looking back. But although the test requires the use of several skills, a comparison with other tests (where you don't have to write anything down, for instance, or where there's no code that can be memorised) has demonstrated that it primarily measures processing speed. An interesting detail: in many studies where a big difference between younger and older people was found on this test, writing speed did not differ between age groups. There are also indications that both groups use the same strategy: repeatedly looking back at the code before filling in the box. The main difference lies in the time each group needs to look at the symbols and digits and then turn that information into action.

A large number of studies have shown that of all the cognitive functions, speed of processing is the one that declines most with age. This has a number of practical implications. Older people who still process information reasonably quickly can continue living independently for longer. And, as Timothy Salthouse suggested, researchers have found that a decline in processing speed can explain a large proportion of the decline in other cognitive functions. In one study carried out at Brunel University in London, a number of tests were administered to a group of people with an average age of 23 and another group with an average age of 68. The capacities tested included memory, executive functions (ability to plan and predict), and speed of processing. The results showed that the older subjects performed less well than their younger counterparts on the memory test, but when the difference in speed of processing was taken into account, they performed equally well. This was not the case when the researchers took differences in other cognitive functions into account, which confirms earlier conclusions that how fast a person can think may explain much of the difference in memory performance between young and old.

RESERVES

Whether your attitude is positive or not, episodic memory, working memory, executive functions, and speed of processing all decline with age. So is growing old nothing but misery? Definitely not. As we will see, there are a number of areas in which older people excel. I'll discuss just two of them here: general knowledge and vocabulary, both standard items on intelligence tests. General-knowledge tests include questions such as 'Who is the president of France?' or 'What is the highest mountain in the world?'. In vocabulary tests, subjects have to explain in their own words the meaning of another word—that is, to play the role of the dictionary. Thus, they might be asked what 'elite' or 'eloquent' means. The words are carefully selected on the basis of a reliable random sample of the population, enabling researchers to establish whether someone is performing above or below average. Vocabulary is crucial to language skills.

Older people perform better in general-knowledge and vocabulary tests. The ability to read and understand narrative texts does not deteriorate significantly as we age, provided there are no problems with eyesight or concentration. Although reading and understanding non-fiction is more difficult for older people because it relies more heavily on working memory, in most cases this only becomes evident after the age of 70.

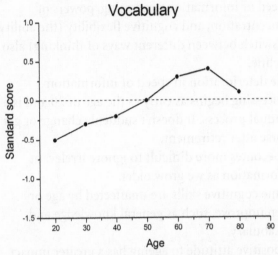

Vocabulary

FIGURE 8. *Vocabulary (and general knowledge) increase until we enter our seventies.*

In psychology, the skills that are not affected or even improve as we age are known as 'crystallised intelligence'. They represent the accumulation of knowledge and abilities throughout a person's life. General knowledge and vocabulary are important components of this kind of intelligence. Functions such as working memory, executive functions, and speed of processing, which do decline with age, are known as 'fluid intelligence'. These are the skills needed to solve new problems, independent of knowledge acquired earlier on. Can fluid intelligence be improved? This is an important question. If this is the case, it may be possible to counteract or compensate for age-related decline. We'll look at this in more detail in chapter 6.

- The ability to retain new information declines as we age.
- Speed of information processing, powers of concentration, and cognitive flexibility (the ability to switch between different ways of thinking) also decline.
- The deterioration in speed of information processing begins as early as the age of 20 and is a gradual process. It doesn't suddenly change or get worse after retirement.
- It becomes more difficult to ignore irrelevant information as we grow older.
- Some cognitive skills are unaffected by age or even improve, such as general knowledge and vocabulary.
- A positive attitude to ageing has a greater impact on health than physical activity, smoking, or obesity.

2

A Calm Disposition:

why older people are more emotionally stable

In January 1990, the British author Roald Dahl was holidaying in Jamaica with his wife, daughter, and granddaughter. The 73-year-old writer had suffered health problems in recent years: he had had two bowel operations, and his eyesight was not good. But according to his biographer, Donald Sturrock, even though Dahl walked like an 'arthritic giraffe', he was calm and cheerful. The pleasant warmth of the Jamaican sun increased his sense of relaxation. Dahl lived very much in the present and enjoyed the company of the people he loved. Evidently fond of metaphors drawn from the animal world, Sturrock described his subject as 'a grizzled lion surrounded by three glamorous lionesses'. But it appears that, after an extremely busy life, Dahl had achieved a measure of peace even before the holiday in Jamaica. 'A kind of serenity settles upon you like a warm mist,' he wrote. 'The real struggle is over. Every movement becomes slower. You have all the time in the world. There is no rush. The never-ending fight to achieve something excellent has ended.'[1]

But old age is not accompanied by serenity for everyone. According to the American psychologist Erik Erikson (1902–1994), there are eight stages of development in a person's life. The last stage begins around our 65th year. He dubbed this stage 'ego-integrity versus despair'. He meant that, at this point, people reflect on the past and develop feelings of acceptance and contentment if they feel they have led a happy, productive life. Erikson called a correspondence between how you wanted to live and how you did in fact live 'ego-integrity'. But if you are unhappy with your life's course and your own role in that, you are likely to experience feelings of despair. Of course, for many people the situation won't be so black-and-white. Although Erikson formulated his theory in middle age, before he knew how it would apply to himself, he became a shining example of ego-integrity. He died at 91, having spent the last years of his life with his wife, Joan, happy and contented with what he had achieved. They were often seen walking hand in hand and kissing openly in public.

Both Dahl and Erikson talked about growing peace and contentment as the years go by. Does emotional stability increase as we age?

PERSONALITY CHANGES

To a large extent, emotional stability is a personality trait. Some people have very little of it: they are easily upset. If they break a cup in the kitchen, they react as if a catastrophe has occurred. Other minor setbacks, such as missing the bus or forgetting their wallet when they go shopping, make them equally distressed. Criticism makes them angry or uncertain. They are often anxious or tense, worrying about approaching disaster or what other people think of them. They are easily embarrassed, fret more, and are more fearful. In short, negative feelings predominate. Psychologists call this 'neuroticism' or 'emotional instability'. Around 15 per cent of the population scores high on the standard measures of neuroticism. Neurotic people are sensitive, which is not purely negative. But neuroticism has clear drawbacks: people with this personality trait are more likely to develop psychological problems such as depression and anxiety disorders.

'Older people are often ill-tempered and irritable.' At least, that's the stereotype. In fact, neuroticism declines in older people. They are more stable and less easily upset than younger people, who are more likely to panic in a disaster. Various studies, including research done at Stanford University in California, show that we develop greater emotional stability over the course of our lives. People aged 60 and above feel happier than people between 20 and 40. They experience fewer negative feelings. And although such feelings increase again around the age of 70, they never again reach the levels experienced by younger adults. The increase in negative emotions after 70 is possibly due to growing physical limitations and the gradual loss of one's peers. But it may also be due to changes in the prefrontal cortex (an area at the front of the brain), which regulates emotions. Nevertheless, older people are more emotionally balanced, and they are better able to deal with difficult situations involving others because they can put negative feelings into perspective. Older people are thus less neurotic, as evidenced by their lower risk of psychological problems. Such disorders mostly manifest themselves before the age of 40. When Dutch journalist Henk Spaan was asked if he had benefited from psychotherapy, which had been so popular in left-wing intellectual circles in the 1970s, he answered, 'Oh, you know, growing older helps too.'

Yet another baseless stereotype is that older people complain more. A nationwide survey in the US revealed that it is in fact young or middle-aged people who complain the most about minor pains or ailments, despite the fact that seniors on the whole have more health problems.

Another important personality trait that changes as people age is whether they are 'good-natured' or 'friendly' (or, as psychologists put it, whether they are 'altruistic') — in other words, whether a person is friendly and co-operative in their dealings with others. Tolerance and helpfulness are elements of this. People who score well on these indicators are nice to be with. They tend not to be obstructive or negative. This trait, like neuroticism, can be measured with a questionnaire. It is common to cultures all over the world, although some individuals obviously possess it to a much greater degree than others. Many score better on this character-istic as they grow older.

We'll look later at the question of why older people are more stable and stress-resistant, and how the brain is responsible for this. But we will also see how, contrary to the general rule, depression and apathy can increase at an advanced age.

LIVING IN THE HERE AND NOW

Researchers have put forward a range of arguments to explain why older people are more emotionally stable.

For example, older people may have experienced many difficult situations in the course of their lives and may therefore be less upset by distressing events. Another interesting hypothesis suggests that, in comparison to their younger counterparts, older people are less oriented towards the future and less uncertain about all the things that still have to be accomplished. In other words, they tend to live in the here and now. That creates a more relaxed situation and gives them the chance to appreciate all the little things that make life enjoyable. Younger people are sometimes prepared to accept negative experiences (a demanding and unpleasant boss, for instance) because they serve a longer-term goal (such as a promotion to a better position). Such considerations are less relevant to older people, so they are less prepared to accept negative situations.

Yet another explanation is that seniors are less demanding and more easily satisfied than the young. This could be because they are wiser and more realistic, but it could also be that their standards are lower: they take their age-related limitations into account. Results from a number of studies would seem to support this hypothesis. Furthermore, it's perfectly possible that older people have a better understanding of what makes them happy and so make better choices in their daily lives. Research has shown that they are less likely than younger people to regret a purchase that they have made. This is partly to do with their being less impulsive. Studies have also looked at the way people remember information about products they wish to buy: older people recall the positive aspects of products more than younger people do. Other memory tests (where subjects have to memorise a list of words or images) have shown that while older people perform less well when they have to retain neutral, random words, they often do just as well as their younger counterparts if the list contains meaningful word combinations with an emotional connection. For example, younger people remember 'street, sail, mouse' better than their seniors, but not 'summer, party, music'.

Susan Charles and Laura Carstensen of the University of California put forward an intriguing theory. On the basis of research, they suggest that the decline in speed of information processing as people age can benefit social and emotional functioning. In other words, it can sometimes be a good thing not to react too quickly. If someone takes a sly dig at you, a considered reaction is often better than an immediate, unrestrained response.

And, finally, an important explanation of why older people experience fewer negative emotions is that they have had more practice in dealing with them. In effect, they have become experts over the course of their lives. This point is worth investigating in more detail, but first we have know more about emotions in general.

THE IMPORTANCE OF EMOTIONS

What would life be like if we always felt the same, if our mood never changed? Probably boring and colourless. What's more, without emotions we would often be unable to make decisions. In fact, it might even be dangerous. We have emotions for a reason: they identify what is important. They help us make difficult decisions, such as buying a house. In that situation, people often say, 'I just had a good feeling about it.' The emotions of shock and fear make us run away if we encounter a dangerous animal, or step back if we start to cross the road when a car is coming. The feeling of being 'in love' is essential to romantic relationships. Anger can be a signal that we no longer accept a situation and demand a change.

Emotions also play a crucial role in our social life. Social psychologist Gerben van Kleef says that without emotions we cannot know what motivates others, what they want from us, and what we can expect from them.[2] But emotions have a darker side. They can get in our way and have a negative effect on our lives. Just think of someone who goes through a long period of low spirits: this can easily turn into depression. The important thing is that we learn to deal with our emotions. Someone who can identify, name, and understand their emotions is more balanced, healthy, and supportive of others than someone who can't.

In 1903, one of the founding figures of modern psychology, the German professor Wilhelm Wundt (1832–1920), proposed a classification of emotions according to three dimensions, a system that is still valid today. First, emotions can be described as falling within the range 'pleasurable to unpleasurable'. Obviously, happiness is pleasurable and sadness unpleasurable. But sadness has degrees of intensity, and intense sadness is more unpleasant than mild sadness. The second dimension is 'attention–rejection'. If you feel afraid, you want to avoid (or reject) something. If you're angry, you may want to tackle (or pay attention to) the object of your anger. Wundt's third dimension was the degree of physical arousal associated with the emotion ('strain or relaxation'). Arousal may manifest itself in an increase in heart rate, blood pressure, respiratory rate, and sweating. This is in itself very important, as it is the way the body prepares itself for action. But if it goes on for too long, as it does in people suffering from depression or anxiety disorders, it can damage your health. So how we deal with emotions is of vital importance. Though people often say they are 'overcome' by emotions, we are in fact capable of influencing their intensity and nature.

AGEING BRAINS AND EMOTIONS

People differ in the way they deal with emotions. Many seldom stop to think about how they are feeling, nor do they talk much about their feelings. Yet it is important to be able to do this if you are to be comfortable with yourself. Adults, unlike children, are expected to be able to control their emotions: allowing your boss to see your irritation when he saddles you with yet another task could cost you your job.

Most people adopt one of two ways of dealing with negative emotions: either repressing the emotion or reinterpreting the situation so that the emotion becomes less negative. Repressing emotion involves putting a brake on emotional expression — by controlling your facial muscles, tone of voice, or body language, you hide the fact that you are angry, afraid, or sad. In other words, emotional repression means behaving as if everything is normal while you are shaking with fear, torn by grief, eaten up by jealousy, or boiling with anger inside. And sometimes this can be useful — for example, in negotiations where you have to prevent the other party from seeing how much you want something.

Reinterpreting the situation is a very different strategy. Here, you focus on the cause of the emotion and try to see it in a different light. For instance, if your partner makes an unkind remark, you can take it personally and start an argument. Or you can assume they're not feeling too good or have had a bad day at work. In that case, you judge the remark less harshly, and the emotion you feel is less negative.

Reinterpretation involves seeing the nuances in a situation or focusing on the circumstances that have led to an unpleasant experience. Someone once spilled hot coffee on my leg in the train. I was very irritated by what I saw as clumsiness. I didn't say anything because it's not in my nature to have a go at complete strangers, but repressing my irritation didn't diminish it. Yet when I remembered that the train had jolted suddenly and the person in question probably couldn't help spilling the coffee, I felt much less annoyed.

A lot of psychological research has shown that reinterpretation is a better way of dealing with emotions than repression. It makes people feel more positive, and slows their heart rate and blood pressure. If you repress the emotion, your heart rate and blood pressure remain high. The same goes for cortisol, the hormone that is released in response to emotion or stress. If heart rate, blood pressure, and cortisol levels remain elevated for a long period of time, it can damage your health.[3] So reinterpretation is good not only for the mind but also for the body. People who reinterpret rather than repress emotions suffer less from heart and vascular problems and from age-related cognitive decline. In addition, they are more emotionally balanced and have better social relationships.

What happens in the brain when we feel emotion? First, the amygdala becomes active. This is a small structure located deep in the temporal lobe of the brain (see figure 9), and it is crucial to experiencing emotion. If we experience negative emotions over a long period, the amygdala remains active throughout that time.

Its activity is linked to raised heart rate, blood pressure, and cortisol expression. When the emotion subsides, the amygdala becomes less active. If a person takes a different perspective on a situation, activity in the frontal part of the brain increases, slowing down the amygdala and producing a feeling of greater calmness.

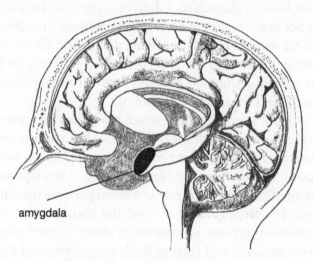

FIGURE 9. *Location of the amygdala in the brain*

How does this work in older people? Studies using brain scanners have revealed a pattern of brain activity which supports the finding that older people are in general good at regulating their emotions. In people aged around 70, the frontal part of the brain (known as the prefrontal cortex) is more active during the reinterpretation of an emotional situation than in younger adults (around 25). The prefrontal cortex controls the activity of many other areas of the brain, and plays an important role in influencing emotions. Though the prefrontal cortex is more active in older people, the activity of the amygdala decreases with age. So it seems that older people use the frontal part of their brain more than younger people. As a result, they are better at calming the amygdala. This is probably because of their greater experience in dealing with emotions. Research into optimism and the brain shows that a positive attitude is accompanied by activity in the parts of the brain that regulate emotion (in the prefrontal cortex). This could explain why older people with a positive attitude are better able to deal with stress and emotion.

An Australian study involving 242 people between the ages of 12 and 79 showed that emotional stability is directly related to age: the older you are, the more stable. Eighty subjects were shown pictures of fearful or happy faces while their brain responses were tracked with MRI scans. Older people had more activity in their frontal lobes when looking at fearful faces, but less when confronted with happy faces. Since these areas in the brain play a part in regulating emotions, the researchers concluded that older people seem better able to control negative emotions, but in fact exert less control when it comes to positive emotions. This suggests they are more open to such emotions.

DARK MOMENTS IN LATER LIFE

Despite the increased ability of older people to cope with emotions, depression can still strike. On visits to her mother, Connie (aged 77), Karen (aged 48) noticed that Connie was often in low spirits. On the phone, she was talking less and less. Karen's father had died five years before, and though her mother seemed to have coped reasonably well, the negative emotions resurfaced when she lost her best friend as well. But she wasn't constantly down. Karen thought a fall in the kitchen had triggered her mother's (current) state of persistent gloom. Though the doctor said Connie had only bruised her right arm and shoulder slightly, she suffered considerable pain in the first few days after the accident and needed help in doing her hair. She often sat staring into the distance and had little appetite. A carer asked her if she might have a hobby she could take up again. But Connie dismissed all her suggestions. She had always enjoyed making her own greeting cards, but had no interest in doing that now. Then the doctor prescribed antidepressants — Karen hoped that they would work, and that her mother would return to normal. After two weeks on the medication, Connie's depression did indeed lift. Now, Karen is doing everything to ensure that her mother feels supported, and organises regular visits from family and friends.

Feeling depressed is slightly more common in people over 70, as in the example of Karen's mother. In Europe, 12 per cent of people aged 70 or older suffer from feelings of depression, compared with 9 per cent of middle-aged people. Yet serious depression is rarer in old age. Of course, it's important to distinguish between feeling depressed and a diagnosis of clinical depression. Feelings of depression are milder; people suffering from them are often described as 'being in low spirits' or 'apathetic'. They feel flat; have lost interest in their daily activities, which seem pointless; and focus on the negative aspects of life or of their own personality. Interest in social relationships and hobbies declines; and people can be irritable, introverted, and inactive. If this continues for longer than two weeks, and clearly gets in the way of normal functioning, then clinical depression is the probable diagnosis. This can take serious forms: lying in bed all day, hardly speaking, having recurrent thoughts about death or even suicide. Professional help is essential.

The slight increase in depression among older people usually has nothing to do with being less able to deal with negative emotions or to change them, since that ability improves with age. So what is the reason? That is not yet entirely clear, but there are a number of possible explanations. First, there is, of course, the fact that as the years go by, we are confronted more frequently with the loss of loved ones, as was the case with Karen's mother. Second, the decline in physical and mental activity can feed depression. Obviously, problems such as having difficulty walking or being prone to falls do not make people happier, and a number of studies have linked these problems with depression. In other words, people with physical ailments are more likely to feel depressed. In our example, Connie may have realised after her fall how vulnerable an ageing body is. Perhaps she saw herself shuffling along with a walking frame in the near future. Third, an older person may look back on his or her life and experience feelings of dissatisfaction, feelings that Erikson called 'despair'. He suggested that those who are dissatisfied with their lives or the role they have played are likely to experience despair. This may explain the finding that people from lower social and economic classes tend to suffer more from feelings of depression, though this is not connected to poorer access to healthcare and other services. Psychotherapy and pastoral care might help older people to cope with feelings of despair and accept their lives for what they are.

Finally, certain changes in the brain may explain depression. One is the shrinking of neurons (nerve cells) in areas that are important for processing negative emotions. Another is arteriosclerosis, or hardening of the arteries, which reduces the supply of oxygen to some areas of the brain, causing transient ischaemic attacks (minor strokes) and damaging brain tissue. Depression can also accompany early dementia, where it is probably the result of brain-tissue damage caused by Alzheimer's disease. A correct diagnosis is essential: depression in older people often improves after treatment.

SOCIAL RELATIONSHIPS

Our emotional life is closely linked to the way we function socially. Social contacts are important for everyone, particularly as we grow older and many of our usual daily interactions (with workmates, for instance) come to an end. People are social beings: most of us find it unpleasant to be alone for long periods of time. Older people often appreciate social contact more than their younger counterparts. They can be more aware of the positive role played by others. In one study, video recordings were made of conversations between couples, young and old, on a subject they disagreed about. Afterwards, the recorded conversations were carefully analysed. When asked about their partner's attitude, older couples were much more positive than younger couples. An interesting finding was that older couples were even more positive about their partner than independent assessors. Positive social contacts are good for our emotional wellbeing and for our mental faculties. After a stroke, for instance, memory and concentration recover faster in people with good social support. Research shows that the cognitive faculties of older men who have lived alone for a number of years decline faster than those of men who live with someone else.

Not only are social relationships good for older people, but they also have a better insight into these relationships. Researchers at the University of Michigan found that people over 60 had a better understanding of conflicts between groups, and were better at resolving such conflicts, than people under 60. They compared three groups of subjects: people aged between 25 and 40, between 41 and 60, and over the age of 60 (with an average age of 70). The subjects were asked to think about a number of fictional situations involving a clear conflict between groups of people. One example concerned immigration. Attracted by strong economic growth, a large number of people migrate from Kyrgyzstan to Tajikistan. They try to preserve their own way of life and customs, but the Tajiks want them to assimilate fully and abandon their traditions. The subjects in the experiment were asked two questions: 'What do you think will happen after they migrate?' and 'Why do you think it will happen in this way?' The answers were assessed by a panel of experts on the basis of criteria drawn from the literature on insight and wisdom, and approved by professional advisers and scientists in this field. To demonstrate social insight, the answers had to show evidence of one or more of the following:
- ability to empathise with the perspective of the people involved in the conflict
- awareness that the situation can change
- flexibility in predicting various possible developments in the conflict
- recognition that many things are uncertain and our knowledge is limited

- willingness to search for a solution or to compromise.

The answers given by the subjects in the study revealed that older participants scored highest on social insight. In response to the immigration situation, one older subject commented: 'While allowing the people from Kyrgyzstan to keep their customs, the Tajiks could also take steps to encourage them to integrate. In other words, there could be ways — for instance, by combining certain customs — to create more unity in the country without causing people to lose their identity.'[4] This shows a definite willingness to compromise. Compare this with the response from a younger participant: 'I'm sure that each culture will keep its own customs. It's hardly likely that a person will change their way of life simply because they've moved to another place.'

Another feature of social insight is the ability to adopt the perspective of another person. This also proved more common among the older participants in the study. One example is the following response of an older subject, again with reference to the dilemma confronting people from Kyrgyzstan who move to Tajikistan: 'People always integrate in the end, but it can take several generations. There will be reciprocal influences, but people in the countries receiving immigrants see it from their own perspective, which is that the newcomers are changing the country. There is, of course, a different perspective: immigrants themselves may be concerned that their children are different from how they would have been if they had grown up in their country of origin. But people in the host country don't necessarily understand this.' The younger participants provided fewer statements that showed an ability to shift perspective. One example: 'What will happen is that a situation will develop like in the United States: people who are only interested in the economy will want to promote immigration, while the traditionalists want legislation to be passed prohibiting the use of any language except that of the country itself. Politicians will come under severe pressure, and it is likely that a left- or right-wing leader will emerge who will fight against immigration.'

Because the researchers consistently found more statements from the older subjects that indicated the presence of social insight, they concluded that older people were better at reasoning about complex social situations. They therefore advise involving older people in legal proceedings, counselling, and negotiations in situations where conflicts between social groups have to be resolved.

IMPORTANT INSIGHTS
- As we grow older, we are less troubled by negative emotions. Older people are more stable and can cope better with their feelings.
- People aged 60 and older are happier than those between 20 and 40.
- Older people are likely to be nicer than 20-year-olds.
- People aged over 70 can suffer from feelings of depression, more than the middle-aged. But serious depression is much less common in older people.
- As a result of their life experience, older people have more insight into complex social situations.

3

Grey Cells:

the anatomy of the ageing brain

The dark interior of the red-brick building at number 37 on Madrid's sunny Avenida del Doctor Arce is home to the Cajal Institute, a neuroscience research centre set up early in the 20th century. It is named after its first director, Nobel Prize winner Santiago Ramón y Cajal, famous for his precise drawings of brain cells (neurons) and the methods he developed to accurately reproduce the cell structure. He was also a pioneer when it came to understanding what happens to neurons as we age.

Cajal's oak desk still stands in the Institute's library. Here, he did his painstaking work, producing countless anatomical drawings. It is quite an experience to see the originals: not only was Cajal a scientific genius, but he also had artistic talent. He was able to depict the neurons and their dendrites (branch-like projections) in such detail because he used chemical substances to colour brain tissue so that its structure was clearly visible under the microscope. Cajal used thin slices of tissue from the brains of cadavers, along with colouring agents developed shortly before by the Italian doctor and scientist Camilo Golgi.

Through his accurate observations, Cajal disproved the prevailing theory (which Golgi also subscribed to) that the brain consists of a network of cells connected by threads, like a spider's web. Cajal discovered that the dendrites of nerve cells do not touch, but are separated by a tiny gap known as the synaptic cleft. Neurons communicate with one another not through a physical connection but through chemical messengers (neurotransmitters), which traverse the gap between neurons and attach themselves to receptors in the neuron's cell wall, like a key in a lock. This is how they transmit signals, and each substance has its own receptor.

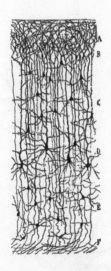

FIGURE 10. *Drawing of neurons by Nobel Prize winner and neuroscientist Santiago Ramón y Cajal (1852–1934).*

Cajal's important discoveries led to much greater knowledge about how the brain works. Communication between neurons is the basis for functions such as memory, concentration, and thinking skills. If we want to know what changes in our brains as we get older, Cajal's drawings show us the main elements of that process. Figure 10 is one of his drawings. The black dots are the neurons, or 'control centres'. The dendrites are the tiny projections emerging from the neuron wall. The longer threads are called axons: they carry signals from one neuron to another. Figure 11 shows a simplified version of two neurons that are in contact with each other.

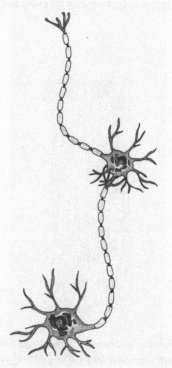

FIGURE 11. *Neurons and their dendrites.*

As figure 11 clearly shows, a neuron has one long projection (the axon), and lots of dendrites. Axons are enclosed in a fatty sheath, known as myelin, which increases the speed at which signals are transmitted. The areas of the brain where many axons come together are known as 'white matter', because of the colour of the myelin layer. 'Grey matter' is mainly at the outer edge of the brain, the cerebral cortex, where the cell bodies of the neurons are. White and grey matter are clearly visible on an MRI scan.

As I explained, neurons use neurotransmitters to send signals to other neurons. The neurotransmitters either increase or reduce the chance that the neuron in question 'depolarises', which means that an electrical signal is transmitted from the cell nucleus to the end of the axon. That process can activate another neuron, and then another, eventually forming a full circuit. A circuit consists of groups of neurons (hundreds of thousands of them), sometimes from different areas of the brain, which are connected to one another and work together, for instance when we speak or read. Cajal demonstrated that neurons form the building blocks of our brains

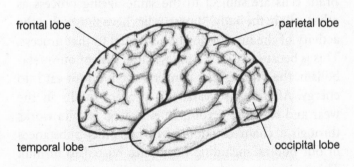

FIGURE 12. *The brain is divided into four lobes: the frontal, temporal, parietal, and occipital lobes.*

and are the basis for all brain activity. The brain has around 100 billion neurons distributed over four large areas: the frontal, temporal, parietal, and occipital lobes (see figure 12).

So what happens in our brains as we grow older? We know of course that something changes, since our mental capacities change, as we saw before, and our mental capacities are inextricably bound up with brain anatomy and function. But what exactly changes? Do neurons die? Does the form or composition of the neuron cell nucleus change? Or is it to do with the projections (some might say the 'wiring') or the fatty sheath around the axons, which normally speeds up brain activity? And which area of the brain suffers the most because of those changes? There are enough questions here to warrant a closer look at the anatomy of the brain. But before we do that, we have to know more about general ageing processes in our bodies.

CHANGES AT CELL LEVEL

Brain cells are subject to the same ageing process as other cells in the body. Many studies have shown that the activity of the insulin hormone is vital to that process. This is because insulin is an essential part of our metabolism, the process that converts the food we eat into energy. After many years, this process results in the wear and tear that accompanies ageing. Insulin works through a chain reaction involving other substances in our bodies, including hormones related to insulin. This is known as the glucose–insulin system. People whose glucose–insulin system is not highly active tend to live longer. Its activity is related to how many calories you consume — in other words, people who eat little, but consume a healthy portion of food every day, live longer. This idea is not new: it was proposed as far back as the 2nd century CE by the Greek physician Galen as an effective prescription for a longer life. The 16th-century writer Alvise Cornaro was also a firm believer in the theory. Since then, countless studies have confirmed this view, although most of them used animal subjects such as mice. Reduced calorie intake is also a feature of the Okinawa diet, which I will discuss later.

Thus, eating less means you can live a longer and healthier life. For some people, this is an automatic process: they are genetically programmed to eat in moderation. Although it's not entirely clear how eating less leads to a longer life, biologists believe that when times are hard and an organism gets very little food, a physiological reaction is launched. This protects cells through reduced activity in the insulin-signalling pathways and so increases survival chances. Or to put it another way: a slower metabolic rate means less wear and tear.[1]

Connected to this is another mechanism that plays an important role in ageing: oxidative stress. This occurs when chemically reactive molecules containing oxygen are produced in greater amounts than normal, which can result in cell damage. Some of these molecules are free radicals, a natural by-product of breathing — that is, we all have them.

Oxidative stress is a cumulative process associated with ageing, and it is particularly harmful to DNA, the genetic information contained in the cell. Although it was long thought that oxidative stress is always harmful to the cells, it now seems likely that a degree of stress can do no harm — only very high levels are damaging.

Eating a lot increases the metabolic rate and therefore oxidative stress, while eating just a little has a protective effect.

There are many other things that influence the ageing of our cells besides the glucose–insulin system and oxidative stress. One example is the mitochondrial damage that builds up over the years. Mitochondria are structures essential to cells, including neurons, because they play an important role in metabolism. Another influential factor is the weakening of our lungs and hearts as we age. As our blood vessels become harder and thinner, fewer nutrients and less oxygen reach the brain. This can lead to hypoxaemia — reduced blood oxygen levels — or, in the worst-case scenario, anoxia, which is a complete lack of oxygen. This can result in cell damage and even cell death, and may lead directly to reduced brain function.

A BRAIN AGED 115

Only 20 years ago, scientists thought that as we grew older our brain cells started to die, a process that began at birth and sped up after the age of 70. New research has shown, however, that most brain cells remain reasonably intact until we die (as long as there is no brain disease present). This was certainly the case with 115-year-old Hendrikje van Andel, whose brain was studied by neuropathologists at the University Medical Centre Groningen (UMCG) immediately after her death.

Reaching the age of 100 is a special event for anyone who lives that long, but particularly for Hendrikje. When she was born in 1890, she weighed only 1.6 kilograms and her chances of survival were low. As a child, her health was poor: she became ill after her first day at school, and her parents decided that she should stay at home. Her father was a teacher and he home-schooled her; she went on to become a teacher herself.

Until the age of 105, Hendrikje lived independently. After that, her sight deteriorated so much that she had to move to a retirement home. But mentally she was still alert. She listened to the radio every day and knew what was going on in politics and sport. At the age of 82, not knowing how long she would live, she had decided to donate her body to medical science after her death. When she was 111, she contacted UMCG to ask them whether such an old and fragile body was still useful to them. So the scientists decided to pay her a visit. They explained to her that her body was of great significance for research and asked if she was prepared to undergo a number of cognitive tests. She was glad to be able to do something for science, and undertook two sets of neuropsychological tests: once when she was 112, and then again when she was nearly 114. These showed that she had an exceptionally good memory for her age — in fact, she could recall stories slightly better than the average 70-year-old. She was also able to focus her attention and recognise objects by touch. The second set of tests showed that she had slightly more difficulty with questions that called on her working memory and with reasoning, but she had no clear cognitive disorders and no sign of Alzheimer's. This was confirmed by the post-mortem examination. Hendrikje's brain tissue was remarkably intact — unlike in people with Alzheimer's, there was little build-up of protein and there was hardly any sign of shrunken or dead neurons. Nor was there any significant hardening of the arteries. In fact, the cause of her death had nothing to do with her brain: she died of undiagnosed stomach cancer.

DNA research has since shown that genetics probably played an important role in keeping her brain healthy as she grew older: she proved to have the 'good' variant of a whole series of genes that are linked to Alzheimer's. Her mother lived to the age of 100, which also indicates an inherited element. According to the researchers in Groningen, the case of Hendrikje van Andel shows that reaching an extremely advanced age is not automatically accompanied by substantial deterioration in the brain or by brain disease. Nevertheless, changes take place even in the brains of healthy people as they grow older.

CHANGES IN THE BRAIN THAT ACCOMPANY AGEING

Generally speaking, the weight and volume of the brain diminish as we move into old age. The brain grows continuously from infancy until about the age of 21. For much of the previous century, you weren't treated as an adult until you reached the age of 21; and in terms of neuroscience, you might say that 21 was a less random choice of age for adulthood than 18. Between the ages of 20 and 50, brain volume remains fairly constant, although this depends on the individual. After that, it gradually begins to decline (a total of about 10 per cent), and around the 80th year of life you see a substantial reduction in volume. Areas such as the frontal cortex and the hippocampus (figure 13) are more affected by ageing than other parts of the brain. This can be seen in the reduced volume of grey and white matter and in the level of activity in these areas. The frontal cortex is involved in planning and looking ahead, in working memory, and in organising and monitoring our behaviour, while the hippocampus is essential to our long-term memory, particularly information storage (though the frontal cortex, in its turn, is instrumental in retrieving that stored information). So one answer to the question of why older people are more forgetful could be that their hippocampus has shrunk slightly, which means they find it more difficult to retain information.

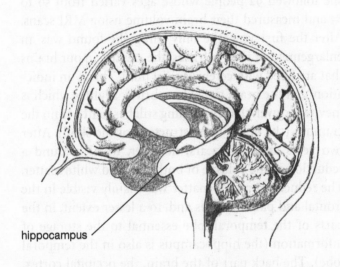

hippocampus

FIGURE 13. *Location of the hippocampus in the brain.*

Large-scale studies using MRI scans to measure brain volume show that the frontal parts of the brain shrink more than the areas at the back. Some of these studies compared a group of older people (aged 60 to 80) with a younger group (20 to 40). This is not an ideal method, since other differences between the groups besides age may play a role. Older people may have different lifestyles, social contacts, or eating habits, all of which could influence the results. From a scientific point of view, a more convincing result can be obtained by studying the same group of people over a number of years, subjecting them to MRI scans every ten years or so, and tracking changes in their brains. This is known as longitudinal research.

The American neurologist Susan Resnick, who is the principal investigator on the Baltimore Longitudinal Study of Ageing, did just such a study. For four years, she followed 92 people whose ages varied from 59 to 85 and measured their brain volume using MRI scans. After the first year, the only thing she found was an enlargement of the ventricles, the cavities in our brains that are filled with cerebrospinal fluid. This is an indication that the brain tissue has shrunk slightly, which is inevitable given that everything still has to fit within the cranium (the upper bony structure of the skull). After two years and four years, however, Resnick found a reduction in the volume of both grey and white matter. The reduction in grey matter was mainly visible in the frontal and parietal lobes and, to a lesser extent, in the parts of the temporal lobe essential to the storage of information (the hippocampus is also in the temporal lobe). The back part of the brain, the occipital cortex, was the least affected.

In contrast, the reduction in white matter was spread over the entire brain. The reduction in white matter is a sign of a decrease in myelin, the fatty sheath surrounding the axons that increases the speed at which signals are transmitted. This could of course be the underlying reason for the diminished processing speed in older people that I described earlier. MRI studies have in fact shown that older people whose white-matter tracts (the 'wiring' of our brains) are no longer intact respond more slowly in neuropsychological tests such as the digit symbol substitution test discussed in chapter 1.[2]

Other studies focused exclusively on changes in white matter as people aged. A common finding in all these studies was that minor damage to white matter, known as white-matter changes, can be seen from around the age of 50. On an MRI scan, these show up as small, bright white spots. The older a person is, the more of these minor injuries can be seen. Although they aren't necessarily serious, they do have a relationship to diminished cognitive function. This was demonstrated in a study carried out by the University of Edinburgh. People with an average age of 78 were invited to participate in a study involving MRI scans of their brains. The unusual feature of this study was that researchers had access to test results of the subjects' mental capacities (such as memory and concentration) at age 11. In 1932, the subjects had taken part in a large-scale regional study, the Scottish Mental Survey, which included tests of reasoning, speed of processing, and memory. After undergoing a brain scan, the same people, now aged 78, took a number of tests measuring cognitive function. The crucial question was what would be a better predictor of their current mental capacities: the test results from when they were children, or the white-matter changes visible now? The answer was remarkable: they were equally good predictors. About 14 per cent of their current performance could be explained on the basis of their childhood scores, while another 14 per cent could be explained by white-matter changes. In other words, if you scored well as a child, there was a greater chance that you would do the same at age 78. But white-matter changes can affect the picture. If you have them, you perform less well on memory and speed-of-processing tests. The individual differences were interesting. If Gareth had a better memory than Mary in 1932, that might well be reversed 67 years later if he had more white-matter damage.

A final point: it is not just the volume of the brain that shrinks; its weight decreases too. On average, the brain becomes 5 to 10 per cent lighter between the ages of 50 and 80. The furrows (sulci) in the cerebral cortex widen, while the ridges (gyri), which contain the grey matter, become narrower. If brain cells do not die in huge numbers as we age, how can the brain's volume and weight decrease? It's likely that this is partly the result of certain brain cells shrinking. Some of the dendrites die off, but not the neurons themselves or the axons. There is also a decrease in the number of synapses, where the passage of signals from one neuron to another occurs.

REDUCED GROWTH IN BRAIN CELLS

Scientists are just like the rest of us: they too can have fixed ideas that they find difficult to shake off. One such idea in neuroscience was the belief that the mature brain no longer produced new cells — that no new neurons were generated once you reached adulthood; they simply died in greater numbers. People were so convinced of this that any indications that new brain cells could be produced was simply disregarded. In the last 20 years, however, the evidence for neurogenesis (the growth of new neurons) has become so overwhelming that no one can ignore it anymore. Neurogenesis primarily takes place in the hippocampus, the structure that is so important to learning and memory. Thousands of neurons are produced on a daily basis in this area, though admittedly most of them die again within a few weeks. The chance that a neuron will survive is linked to learning processes. When you learn something new, such as a foreign language or how to play the flute, it is highly likely that some of the new cells will be used for this. New cells make it easier to learn something new.

Recently, an interesting hypothesis emerged, arguing that ageing in the brain is largely caused by a reduction in neurogenesis. Is there any evidence for this intriguing supposition? There is certainly evidence for the claim that neurogenesis slows down in the older brain — in fact, it diminishes by 80 per cent. Older brains also contain fewer stem cells (cells that are capable of changing into another type of cell). The question is whether this is connected to poorer mental performance, a link that has been shown in mice. In one study, substances naturally produced in the bodies of older mice were injected into younger mice, causing neurogenesis to slow down.[3] Consequently, the younger mice were less able to find their way through a maze. Research into growth factors in our bodies that promote neurogenesis show that these factors also improve mental capacities in older people. The effectiveness of these growth factors has therefore not diminished — they still play a part in creating new cells or repairing damaged cells — but they have decreased in number. We'll return to this later.

THE PASA PATTERN

For the non-specialist, the titles of research articles are usually fairly dry and boring. But now and again there's a little bit of humour. Take for example the article published in 2008 by an American research group led by neuropsychologist Roberto Cabeza — it was entitled 'Qué PASA? The posterior–anterior shift in aging'. Of course, you have to know that in Spanish-speaking countries, *que pasa* is almost invariably used as an opening remark when people meet. A good English translation would be 'What's up?' So what's up with brain activity in older people? As the title of the article shows, PASA stands for the posterior–anterior shift in ageing, the phenomenon by which less activity in the back of the brain is seen in older people than in their younger peers, while there is correspondingly more activity in the front of the brain. In the last 15 years, functional MRI studies into ageing have established clear changes in brain activity. The numerous results that have been published constantly refer to PASA, the shift from the back to the front of the brain in older people who are performing a task well. In one of the studies, people aged around 70 performed a memory task using words and an observational task using pictures. These are two very different tasks: the first mainly activates the frontal and temporal lobes; the second, the occipital areas at the back. A shift in brain activity towards the front of the brain could be clearly seen in older people.

But what does this mean? The results of a number of studies show that what is happening is a type of compensation. Certain areas of the brain are called upon to compensate as much as possible for a declining ability to perform tasks involving memory, concentration, and co-ordination of thought and action. Cabeza's research shows that in older people undergoing tests while in the MRI scanner, a more noticeable PASA pattern correlates with better performance. In other words, the anterior part of the brain, which we learned earlier is the most affected by ageing in terms of the structure and function of brain tissue, has to draw on all its resources in order to perform well. Calling more on this area helps older people to get the most out of their brains. In many, this process is automatic and unconscious, but it is quite possible that training our mental skills could reinforce this pattern.

PASA is not the only resource older people can draw on. There is another change that is often observed in the older population: a reduction in the asymmetry of brain activity.[4] 'Asymmetry' here means that one side (or hemisphere) of the brain is more active than the other. In the performance of language tasks, for instance, the left side is usually more active than the right. But when listening to emotional music, the right side is more active in most people. Dozens of studies have shown that a reduction in this asymmetry occurs as we age, and this more bilateral pattern of activity mostly takes place in the frontal lobe. Researchers have demonstrated that in younger people, it is primarily the right hemisphere that is active during the performance of visual-attention tasks, while in older subjects both are active. Conversely, younger people performing language tasks (word-based tasks involving working memory) largely called upon the left hemisphere, while in older people the left side was less active and the right side more. Is using the hemispheres more equally also a case of compensation, or is it a more general effect that occurs in ageing — in other words, that the different parts of the brain concern themselves less with specific tasks and are more generally active? More specifically, could an area that in younger people is primarily involved in language processing contribute to visual attention in older people? Smaller areas would thus be less active in specialised tasks, spreading the pattern of activity more generally over the entire brain.

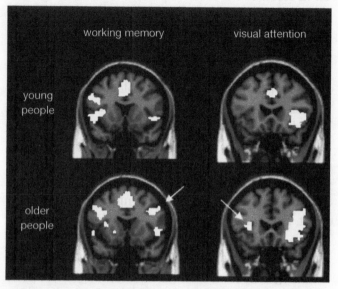

FIGURE 14. *Older people use the right and left hemispheres more equally (see arrows). The right hemisphere contributes to working memory and the left to focusing attention on visual stimuli (e.g. pictures that may appear in different places on a screen).*

Research results point to the existence of compensation: older people who do not display a shift towards greater bilateral activity in the two sides of the brain perform less well in memory tasks than those whose brains show evidence of reduced asymmetry. One study in particular produced clear evidence of this phenomenon. On the basis of brain activity during the recall of information acquired earlier, it distinguished three groups: first, young adults in whom the right frontal cortex was largely activated; second, older people showing the same pattern of activity; and third, older people whose right frontal cortex was less active, but whose left frontal cortex was more strongly activated (that is, who showed reduced asymmetry). Which of the two groups of seniors performed the best? You might imagine it to be the group showing the same pattern as the young adults, who, as expected, performed well. But that was not the case. The group of older people displaying reduced asymmetry performed better than the group showing the same pattern as the young adults. This is strong evidence for reduced symmetry being a form of compensation. More research is needed into compensation, but it is perfectly possible that it involves a reorganisation of neural networks that results in parts of the brain 'co-operating' better with each other.

A TIDY BRAIN

Why do the cognitive skills of some older people continue to be superior to those of others of the same age? At 77, Dutch politician Frits Bolkestein wrote *The Intellectual Temptation*, thought by some critics to be his finest book. Yet there are others in their late seventies who cannot concentrate sufficiently to even read such a book.

To explain this striking difference between people of the same generation, researchers came up with the 'cognitive reserve' hypothesis, according to which our brains have some degree of reserve capacity. Both genetic and environmental factors affect its size. How can we describe this reserve capacity? It is divided into structural capacity and functional capacity. The first is dependent on the amount of intact brain tissue and connections between different areas of the brain (that is, brain structure); the second, with the way different areas of the brain function (that is, brain activity).

A simple example illustrates the difference between the two. Imagine you have a shed full of all kinds of objects and belongings. You decide to put some of them in your attic because the shed has reached full capacity. The attic is your reserve capacity. It is structural because it is a physical space: your reserve capacity is dependent on extra physical structure. But what you could also do is rearrange the objects in the shed to create more space. My wife points this out this every year when I'm packing the car for our summer holidays. So this is comparable to functional reserve capacity — a different strategy for arranging things releases existing space for other uses. Do older people who still function well in cognitive terms have functional reserve capacity? Indeed they do. Making greater use of the areas at the front of the brain (the PASA effect) and of both hemispheres is an example of this. With regard to structural reserve capacity, an undamaged hippocampus can expand that capacity, while the accumulation of protein that accompanies ageing will reduce it. What is also important is whether neurogenesis is taking place. Thus, if you want to keep your cognitive functions in good shape, you need to do something about your brain's reserve capacity. Later on, we'll look at whether that's possible; and if so, how.

IMPORTANT INSIGHTS
- Reaching an extremely advanced age is not always accompanied by a substantial deterioration in the brain.

- Between the ages of 50 and 80, the volume of the brain declines by 10 per cent, and it becomes 5 to 10 per cent lighter.
- The frontal cortex (which plays a role in planning and organising, and in working memory) and the hippocampus (which plays a role in long-term memory and information storage) suffer most from changes associated with ageing.
- A reduction in white matter is associated with slower thought processes.
- New neurons are created even in ageing brains, but a recent study has shown that the rate of neurogenesis declines by 80 per cent.
- Compared with younger people, older people make more use of the front of their brain than the back, to compensate for a decline in function.
- Older people are also more likely to use both the right and left sides of the brain at the same time.

4

Forgetfulness or Dementia?:

where is the line between the two,
and what can you do about it?

At some point, you notice that your grandmother's remarks are becoming very repetitive. And when you mention your sister's new partner, she doesn't know who you mean, though you told her the whole story half an hour ago. This will be a familiar picture to most people with an ageing relative. You think to yourself: *Are these the first signs of dementia?*

Many people, especially once they have turned 60, wonder where the line between forgetfulness and early dementia lies. You forget — for the second time in a few days — where you left the car keys, and have to spend five minutes searching for them. Barbara Strauch, science editor of *The New York Times*, describes in *The Secret Life of the Grown-Up Brain* that she often goes down to the basement of her house to get something, but once she's there she no longer knows what she came for.[1] She looks around her for a while, panicking at the thought that her memory is beginning to fail her. Yet she's only 56. She decides not to give in to this forgetfulness and racks her brains to remember why she came down. Unsuccessfully. It's only once she gets back to the kitchen and sees the empty paper-towel holder that the penny drops. Strauch attributes her memory blackout to her age. But she realises that the problem affects a lot of people and needn't be a sign of pathological decline.

From my own experience, I know that this kind of memory failure can happen regularly from your thirties onwards. I'm not worried about it — yet! The reason is probably that I try to do too many things at once. Research shows that people in their thirties are likely to put these lapses down to their busy lives, while people in their fifties blame them on old age. But the memories of people in the latter age group are mostly in good shape, even if they themselves are not so sure of that.

Nevertheless, a day may come when you or others notice that your mental faculties (memory, concentration, thinking skills) are declining faster than average. Doctors and psychologists call this mild cognitive impairment (or MCI), and there has been an enormous upsurge in research into this issue over the last decade. How do you diagnose MCI? Exactly when should you start to worry about it? Are there demonstrable changes in the brain that herald the onset of dementia? What kinds of dementia are there, and how do they affect the brain?

A DIAGNOSIS OF MCI

Gerard is very concerned about his wife, Esther. She is only 58, but her memory seems to have become much worse over the last few months. She regularly repeats things she said earlier in the day. She forgets routine chores, such as ringing the dentist, or doing the shopping. Esther herself feels she can't trust her memory anymore. Though she's always had a good sense of direction, she's recently got lost in town on more than one occasion. Sometimes she seems confused, and she sleeps badly — even if she takes a sleeping pill, she never gets more than a few hours' sleep. Her GP refers her to a geriatrician. A brain scan reveals no abnormalities. Further tests show that Esther still knows the names of friends and acquaintances, though if she hasn't seen someone for a few weeks she might have difficulty recalling his or her name. She knows what the date is and where she is. And although it requires more effort, she can still do her job. Gerard tells the geriatrician that she has changed in the last six months: from time to time, she appears to have difficulty in following conversations and functioning normally. He's afraid that she is in the first stages of dementia and wonders if there is any medication that could halt the process. The diagnosis is MCI, which can herald the onset of Alzheimer's, though not in everyone. The geriatrician gives Esther and Gerard advice, but also makes it clear that there is no medication that can help.

If you want to find out whether you have signs of MCI, take a look at the following checklist.

– Do you forget things more often than normal?
– Do you forget important things such as appointments, birthdays, and parties (not once or twice, but regularly)?
– Do you regularly lose the thread of a conversation, book, or film?
– Do you increasingly feel unable to take decisions, or to plan the steps needed to complete a task?

- Are you beginning to have difficulty finding the way in a familiar environment?
- Are you becoming more impulsive and less able to assess various situations?
- Have your family and friends noticed these changes?
- Are you suffering from mood swings or any of the following behavioural changes:
  - feelings of depression
  - greater irritation or aggression
  - anxiety
  - apathy?

You don't have to be suffering from all these symptoms. What matters is that cognitive problems are clearly present and have increased in recent months. You're still able to function at home and at work, but people close to you have noticed that you're having memory problems or that you're no longer able to keep a grip on everything. Of course, even if you recognise a number of these signs, that doesn't necessarily mean that you have MCI. That diagnosis can only be made by an expert (a geriatrician, neurologist, or neuropsychologist). But it is advisable to talk to your GP about it, to see whether a referral for further examination is appropriate.

How do people with MCI score on tests that measure cognitive skills? These neuropsychological tests are not always used in diagnosing MCI, though it would be good if they were because they reveal whether those skills are still intact. Specialised memory clinics do use such tests. An exhaustive session charts the various skills: memory, concentration (which neuropsychologists call attention), observation, speed of processing, and the executive functions (the ability to plan and keep an overview, abstract thought, cognitive flexibility, and the capacity to simultaneously process different items of information). You might expect that, on these tests, the performance of people suffering from MCI would lie exactly between that of healthy peers and of dementia patients. Research comparing different groups of people with and without MCI has indeed confirmed this. Yet it is possible for someone with MCI to score normally — or, rather, still score normally — in memory tests. These are people whose memory was previously better than most, so even if their memory decline is above average, they can still perform within the normal range. Take Peter, for example, who at the age of 40 took a 15-word memory test. After 15 minutes, he could still remember 13 of the items read out to him. That is a very good score. However, the test result was not recorded, and Peter himself no longer knows what his score was. At the age of 65, he starts to experience memory problems, which are confirmed by his wife. He does the 15-word test again and can recall seven items after 15 minutes. That is an average score. But his memory has nevertheless declined more than average: while his peers can recall three words fewer than when they were 40, he recalls six words fewer. Unfortunately, most people have no earlier tests for comparison when they first present with memory issues at the GP or clinic. That is why subjective memory problems play an important role in the diagnosis of MCI, particularly if they are confirmed by someone who knows the patient well.

Memory problems are the most prominent symptom of MCI. It's important to keep track of them, since any further deterioration may be a sign of developing dementia. In a test requiring the subject to commit a short story to memory, people in the early stages of dementia almost immediately start to make mistakes in recalling the information. Research has shown that this is mainly the consequence of a lack of concentration while listening to the story. As a result, the information is not properly stored. But memory problems are not the only symptom of MCI. Sufferers' cognitive flexibility is affected, as demonstrated in tests involving the switching of letters and numbers (1 for A, 2 for B, and so on). Compared with healthy peers, people with MCI do less well on this measure.

And what about speed of information processing? As we saw earlier, this is one of the skills that declines the fastest as we age. And indeed it deteriorates faster in many MCI sufferers than in healthy older people. An Australian study showed that poorer performance on the digit symbol substitution test (see figure 6), a common measurement of speed of information processing, was a predictor of the development of MCI four years later. In this study, over 2000 subjects between the ages of 60 and 64 were followed for four years. Of this group, 64 people showed clear signs of cognitive decline. The results of a memory test using words and another test measuring processing speed had some predictive value: a lower score at first measurement was often associated with decline later on.

CHANGES IN THE BRAIN ASSOCIATED WITH MCI

Though memory functions are not located in one specific part of the brain, there are certain areas that are of special importance. The most important is the hippocampus (figure 13) and the surrounding cerebral cortex in the temporal lobe. The hippocampus is a crucial part of the neural network (including the prefrontal cortex) involved in memory. It is not surprising that researchers investigating MCI looked at the structure and activities of this organ. An obvious question in MCI is this: is the hippocampus damaged and its function compromised?

The hippocampus consists of millions of brain cells. MRI scans measuring the quantity of grey matter can tell us if there is a link between reduced volume and the development of Alzheimer's. A recent study combined the results of six longitudinal studies tracking reductions in the size of the hippocampus over a number of years in people with MCI, some of whom developed Alzheimer's and some of whom didn't. The researchers also looked at other brain structures, but the hippocampus and the surrounding cerebral cortex were the only structures that displayed a reliable association with MCI and, later, Alzheimer's. In other words, on the basis of MRI scans it could be said, retrospectively, that a decline in grey matter in the hippocampus correlated with the development of Alzheimer's a number of years later.

The Institute of Psychiatry in London conducted a study of 103 people with MCI that looked at changes in the shape of the hippocampus rather than in its volume. The changes in brain tissue associated with Alzheimer's result in an alteration in the shape of the hippocampus, as measured by a computer program. In 80 per cent of cases, an abnormally shaped hippocampus was accompanied by the development of Alzheimer's one year later.

In addition to grey and white matter, there are other major types of matter in the brain that play a key role in metabolism and the transfer of stimuli. A special MRI technique known as magnetic resonance spectroscopy

(MRS) allows scientists to measure the concentration of such matter. Together with a colleague, I compared all the studies into the differences between older people with MCI and their healthy peers that were performed using MRS. We found that in the hippocampus in particular, the reduction in matter involved substances important for an efficient metabolism.[2] As had been shown earlier, this reduction is much more marked in people with Alzheimer's.

Other studies have revealed that the production of acetylcholine, an important neurotransmitter, declines as we age. Acetylcholine plays a role in learning and memory, but also in activating muscles. In Alzheimer's, the neurons that make acetylcholine are damaged and the levels of the transmitter seriously affected. Medications to combat Alzheimer's therefore aim to activate or mimic the effects of acetylcholine.

Another important change in older brains is the formation of 'tangles' or 'plaques' in brain tissue. As the name implies, tangles are twisted, non-functioning transport proteins (which look like threads and are found in neurons), while plaques consist of insoluble deposits of protein fragments. In Alzheimer's, these proteins are abnormal and damage cerebral function. How this comes about is uncertain, though we know that genetics plays a role.[3]

Figure 15 illustrates the extent to which plaques, tangles, and neuron loss occur in normal ageing, in MCI (as a precursor of Alzheimer's), and in Alzheimer's itself. The illustration at top right shows what the brain of an 80-year-old looks like if that person has few cognitive problems, while the one below on the left belongs to a person suffering from memory problems but not from dementia. The image below on the right shows the brain of a person whose memory problems are so severe that a diagnosis of dementia is appropriate.

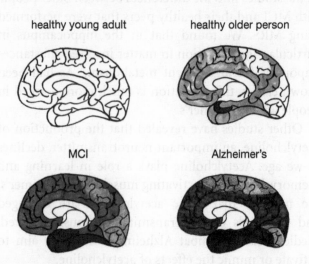

healthy young adult · healthy older person

MCI · Alzheimer's

FIGURE 15. *Tangles and plaques are absent in the brains of young adults (top left); increase slightly in normal ageing (top right); increase further in people with MCI (sometimes a precursor of Alzheimer's), primarily in the temporal lobe (bottom left); and are everywhere in the brains of people suffering from Alzheimer's (bottom right). The darker the colour, the greater the number of tangles and plaques.*

Two things are noteworthy here. First, the more severe the decline in cognitive function, the greater the number of plaques, tangles, and areas with fewer neurons. And second, their location is different: in a person with MCI, the hippocampus is most affected; while in someone with Alzheimer's, a much wider area is involved. Finally, in Alzheimer's there is often inflammation of the brain tissue, which doesn't occur in normal ageing.

An obvious assumption would be that the presence of protein plaques indicates declining cognitive performance. In other words, the more plaques there are, the greater the deterioration in memory and concentration. However, an important question is whether this applies only to dementia sufferers, or if it is also true of people with milder forms of protein build-up, as seen in some otherwise healthy seniors. A problem up to the very recent past was that this kind of protein deposit could only be established through a post-mortem examination. It could not be tracked as people aged. Luckily, a special brain-scanning technique has now been developed to measure protein accumulation.[4] Researchers at the National Institute of Aging in the US used this technique to study 57 healthy people aged around 80. The results of cognitive skills tests from 11 years earlier were available. The scans showed that the older people were, the more protein build-up there was. And the amount deposited was related to the degree of cognitive decline in the 11 years preceding the scan. This study demonstrated that it is not only the accumulation of large amounts of protein characteristic of Alzheimer's that is associated with mental decline. It seems likely that milder forms of protein build-up also have an effect. These milder forms appear in many healthy seniors, and are probably responsible for the less serious decline in skills that is common to the majority of the ageing population.

In the next few years, neuroscientists will attempt to refine even further the analysis of brain scans. The question is whether it's possible to scan the brains of people complaining of cognitive problems to see who is at risk of developing dementia and who isn't. If that were the case, targeted action in the fields of nutrition, exercise, and medication could be taken to prevent people in the high-risk group from developing dementia.

Researchers at the University of Newcastle in Australia recently made the first attempt to find out if it's possible to use scans and other information to determine who is at greater risk of developing Alzheimer's. They combined various test results and used special computer software to see which mix of factors had the greatest predictive power (that is, predicting who would develop Alzheimer's within two years). The researchers examined three factors: the volume of various brain structures as measured by MRI scans, the presence of proteins in the cerebrospinal fluid (obtained through a lumbar puncture) that have been linked to Alzheimer's,[5] and neuropsychological tests (in memory, attention, and executive functions). Values were obtained for each of the three factors from a dozen brain structures (through MRI scans) and a dozen neuropsychological tests. On their own, the tests proved to have the greatest predictive value. But a combination of measurements from each of the three factors predicted the onset of Alzheimer's even better. This included memory tests, the volume of the hippocampus and the surrounding cerebral cortex, and the ratios between certain proteins in the cerebrospinal fluid. The combination provided a correct prediction in 67 per cent of cases.[6] That percentage is not high enough, since in many cases the prediction would be wrong, but it is a step in the right direction. Adding lifestyle factors (diet, exercise, mental activity) to the mix would undoubtedly make the model more successful.

We saw earlier that a reduction in the grey matter in the hippocampus is not a good sign. But what about the *activity* of this organ? Researchers at the University of Berlin scanned the brains of two groups of subjects while they were performing a memory test. The first group had no memory problems; the second reported subjective problems (that is, they themselves thought that their memories were deteriorating, but no objective neuropsychological tests had been performed). The scans showed that the hippocampus was less active in the second group — apparently the hippocampus, the 'memory centre' of the brain, was beginning to slow down. At the same time, the people with memory problems showed more activity in the front part of the brain, particularly on the right side. Despite these differences, the two groups performed equally well on the memory test — there was no decline in actual performance. This confirms the idea that people whose cognitive skills are beginning to decline have a less efficient hippocampus and compensate by using the frontal cortex more.

Another study of brain activity in people with MCI looked at a state in which the brain is 'free', rather than at memory processes. Researchers at the VU Medical Centre in Amsterdam examined patterns of brain activity in people who were lying awake but inactive in the scanner. This is termed 'a resting state'. Of course, even then the brain is busy. Subjects might think about what they did today, or what they're going to do later, or something else. The researchers compared scans of Alzheimer's patients, people with MCI, and healthy volunteers. Then they followed the subjects for three years to see whether the difference in activity predicted the onset of Alzheimer's in the MCI group. Of the 23 participants in that group, seven developed Alzheimer's in the subsequent three years, and 14 remained mentally stable. Of the remaining two, one developed another form of dementia, and the other dropped out of the study. The researchers discovered that the brain scans of those who developed Alzheimer's, made three years earlier when they were outwardly indistinguishable from the other members of the MCI group, showed that two areas in the posterior part of the brain that are important for autobiographic memory and self-awareness worked together less well than in those who did not develop Alzheimer's. Co-operation between different parts of the brain can be measured by checking if they are active in the same way, to the same degree, at the same time. You could compare it to two people dancing together — a folk dance, for instance, where you hold each other's hand. If both co-operate, they move simultaneously. If one of the partners does not co-operate, the pattern is disrupted and there is no harmonious movement. The study revealed that reduced communication between these two areas of the brain was linked to the seriousness of cognitive decline. These findings will probably be used by memory clinics in the future. More specifically, if an increasing reduction in connectivity is demonstrated through, say, six-monthly testing, this could make an important contribution to the early diagnosis of Alzheimer's. In the future, information from such scans could very well be used to diagnose the earliest stages of the disease.

FROM MCI TO ALZHEIMER'S

In May 2011, a controversy raged in the pages of *The New York Times* about the seriousness of MCI. The newspaper was flooded with letters following the publication of an article by American academic Dr Margaret Gullette, in which she argued that many people with MCI managed to cope with their declining cognitive skills. Such people often have a positive outlook, she claimed, and get a lot of support from friends and family. The general tenor of the article was that cognitive impairment is mostly not such a serious matter — a slight decline in memory and thinking skills is a normal part of ageing, and our culture should be less obsessed with perfect mental functioning. A positive outlook keeps you active and prevents further decline. The letters came from indignant relatives of Alzheimer's patients and people suffering from advanced MCI. Having witnessed up close a loved one's unequal struggle with the destructive effects of the disease, they objected in the strongest terms to the view that a positive outlook and support from family are enough to be able to live with cognitive impairment. Neurologists, too, became involved in the debate. They were concerned by Gullette's views because they seemed to trivialise the gravity of the daily problems faced by MCI patients.

The angry relatives were right to say that a positive outlook and lots of help and support, however good, cannot prevent Alzheimer's. Quite apart from that, it is possible that Margaret Gullette and the letter writers were talking about different groups of patients. Various studies have shown that nearly half of the people diagnosed with MCI do not go on to develop Alzheimer's in the following five years. One in seven improves to such an extent that the diagnosis of MCI no longer applies. Gullette was probably thinking of this group, while the letter writers were referring to the other. No one raised this distinction between those who do and those who don't go on to develop dementia in the years following the diagnosis. Nor can this distinction be made when MCI is diagnosed. That is why more research is needed that focuses on exposing the factors that lead to dementia.

Why one person develops dementia and another doesn't has probably to do with a number of factors that reinforce one another. Genetics play a role, as do lifestyle factors (little or no exercise, little mental and social activity, unhealthy diet). MCI that does not progress, where the patient does not develop dementia, is mostly caused by psychological factors such as long-term depression, stress, or burnout. A temporary reduction in brain function caused by vitamin deficiency, low thyroid function, or the side effects of medication can also lead to MCI.

Is everyone at risk of Alzheimer's? Probably not, as we saw with Hendrikje van Andel, whose brain showed no traces of the disease at the age of 115. At the present time, too little is known about the decisive factors in the development of Alzheimer's in people with MCI. Worldwide, a considerable amount of research is devoted to this issue, and it is perfectly possible that in a few years' time we will know much more. Because the disease begins gradually and progresses slowly, it is always preceded by mild cognitive impairment with no disastrous consequences. This means that the first stage is not always recognised as MCI. And there is still no effective treatment. Nevertheless, we do have some idea about steps we can take to exert a favourable influence on the outcome.

WHAT CAN YOU DO ABOUT MCI?

Can we halt the mental decline that accompanies MCI? There is little hard scientific evidence to help us here.

Research has shown that the medication currently available is fairly ineffective. There are two types that can slow down or compensate for the effects of Alzheimer's.[7] One of them raises the concentration of the neurotransmitter acetylcholine, which is crucial to memory processes. The other reduces the effect of the neurotransmitter glutamate on brain cells. Glutamate is the most important chemical messenger in the brain: it ensures that cells activate one another, and thus supports memory and other cognitive abilities. The theory is that in Alzheimer's, glutamate is produced in excessive amounts, damaging brain cells. Hence, the medication blocks its activity. Both medications have been used to treat MCI, but unfortunately neither has proved very successful in preventing Alzheimer's, though one study found a positive effect.

What *has* been proven to have a positive effect is exercise. This was recently established yet again in a study where subjects with MCI (with an average age of 70) followed an intensive exercise program or gentle stretching exercises. The intensive program aimed to achieve aerobic fitness, and the trainer made sure that the participants' heart rate rose substantially during the activity. In the aerobics group, the participants had to keep their heart rate between 75 and 85 per cent of the heart-rate reserve — that is, the difference between a person's maximum heart rate (while running on the treadmill, for example) and resting heart rate. In the control group, doing only gentle stretching, the participants had to keep their heart rate at or below 50 per cent of the heart-rate reserve. The subjects exercised for 45 minutes, four times a week, for six months. Those in the intensive exercise group performed significantly better on a number of tests of cognitive skills compared to the stretching group. The effect was most powerful in the female subjects and primarily in tests designed to measure cognitive flexibility.

We do not know exactly why exercise has a beneficial effect on mental faculties, though better oxygen flow to the brain probably plays a role. There are also indications that intensive exercise promotes the release of growth factors produced by the body, which in turn have a favourable effect on brain cells. We'll come back to this later. In any event, the Roman proverb 'a healthy mind in a healthy body' has lost none of its significance.

But it's not just physical exercise that matters. Remaining mentally and socially active is essential. A study in Chicago of over 1300 subjects between the ages of 70 and 89 showed that older people who use the computer, play games, read books, and engage in creative activities such as quilting or knitting function better mentally than those who don't. Approximately 200 participants with MCI proved to have been much less active in this way in the year before the study than the others. Though this is interesting information, it is important to be cautious about the results of a study such as this, since it reveals no causal link. It doesn't necessarily mean that a less active brain leads to cognitive decline. It could, after all, be the other way round: someone whose mental capacities are declining may engage in less mental activity. And other factors may play a role, too. For example, if you belong to a lower socio-economic class, you may suffer earlier from cognitive deficits, read fewer books, and use the computer less often. Incidentally, the link between mental activity and cognitive decline did not apply to all mental activities. In the Chicago study, the two groups (those with and without MCI) reported reading the newspaper equally often, though it should be noted that many people 'skim' the newspaper, and reading a book demands greater concentration, on the whole.

A study on cognitive training showed that being mentally active is accompanied by an improvement in mental capacities. Such training can take a number of forms. The emphasis often lies on exercises such as memory tests or puzzle-solving. Or the emphasis may be on learning strategies to cope with cognitive tasks. One example is the 'loci method' for training memory. If you want to remember a shopping list, you imagine placing all the items in a particular space (your living room, for example): a carton of milk on the piano stool, half a wholemeal loaf on the table, and a lettuce on the sofa. It seems like a lot of effort, and it is, but that's why it works. You recall what you want to buy as you walk around the supermarket while visualising your living room. Various studies have shown that cognitive skills can be trained in people with MCI, improving their performance in tests of attention, memory, and executive functions. And the effects are long-term. If these skills are retested months later, the participants still perform better than before the training. According to some studies, these effects are 'domain specific', meaning that if you train yourself to concentrate on pictures, your memory will not improve — just your concentration on pictures. Which in turn would suggest that this kind of training has only a limited effect. Other studies, however, have shown that in some kinds of training the improvement is broader than in just the specific task that has been trained. For instance, if a method for training visual attention involves concentrating on small changes in pictures on a screen, this could also improve attention in traffic situations. This doesn't mean that people who undergo this type of training are fully competent to drive. Driving in traffic requires other complex skills, such as anticipating and estimating outcomes (both executive functions), and motor co-ordination. The entire complex of skills can best be tested in a driving simulator in a university neuropsychology department.

Besides these computerised exercises, there are other kinds of training that take place in real life and focus on social participation. They take the form of a daily activity that makes demands on your cognitive skills. Interesting findings emerged from a small-scale study in Baltimore in which older people from disadvantaged social groups (low income, low levels of education) participated in a social and educational program. For 14 hours a week, they worked as teaching assistants with preschool children. Before they started, activity in their frontal cortex was measured through fMRI scans (a form of MRI that measures brain activity through changes in blood flow), and this process was repeated after they had been participating in the program for six months. What was the result? Their frontal cortex was more active, and their executive functions (such as cognitive flexibility) had improved in comparison with a group of people of the same age and background who were on the waiting list for the program.

Older people can probably be encouraged to engage in voluntary social work to a greater extent than happens at present. Rudi Westendorp, former head of geriatrics at the Leiden University Medical Centre, argued this in an interview published in *de Volkskrant*.[8] He pointed out that young people today will probably live to an average age of 90 to 100, while many seniors are still full of life and can make an active contribution to society.

'When she turned 65, my grandmother announced that from now on she was going to take things easy, but she died at the age of 99,' said Westendorp. 'She had to reinvent herself three times over.' Older people of today and in the future will have many more healthy years of life, he claimed. 'One voluntary organisation has calculated that older volunteers have provided billions of hours of free informal care this year. Why don't we convert that into normal, paid work? That way, older people can support themselves for longer and need less pension.' Not to mention the fact that the risk of developing MCI and Alzheimer's is probably lower, or at least their onset is delayed, if people can remain active in this way.

The renowned Mayo Clinic in the US, which specialises in research into ageing and the treatment of Alzheimer's, advises patients to remain physically active, but also to follow a low-fat diet with lots of vegetables and fruit. In addition, it recommends that patients regularly eat oily fish, which contains beneficial omega-3 fatty acids. This may have a positive effect on the heart and thus on the brain. Although these steps can do no harm, it is important to remember that there is substantially less evidence for this view than for the value of physical exercise and mental activity.

IF DEMENTIA DOES STRIKE

Former US president Ronald Reagan was instrumental in opening up discussion on the subject of dementia. In November 1994, he published a letter to the American people announcing that he had been diagnosed with Alzheimer's. 'I have recently been told that I am one of the millions of Americans who will be afflicted with

Alzheimer's disease,' he wrote. 'At the moment, I feel just fine. I intend to live the remainder of the years God gives me on this earth doing the things I have always done.' The letter ended: 'I now begin the journey that will lead me into the sunset of my life. I know that for America there will always be a bright dawn ahead.' According to his son Ron, Reagan was already showing signs of dementia in 1986 (when he was still president).[9] For instance, while flying over the familiar canyons north of Los Angeles, he could no longer summon up their names. The diagnosis was not officially made until 1994, and he died ten years later, at the age of 93.

Age is the biggest risk factor for dementia. While only 1 per cent of 60-year-olds suffer from it, 7 per cent of 70-year-olds and around 30 per cent of 80-year-olds have one form or another. In Australia, 322,000 people are estimated to be dementia sufferers, a figure that can only grow as the population ages. The Australian Institute of Health and Welfare estimates that it will reach 900,000 by 2050. And according to the World Alzheimer Report 2009, 118 million people worldwide will be suffering from the disease by 2050.

The most common forms of dementia are Alzheimer's disease, vascular dementia, frontotemporal dementia, and Parkinson's disease. Alzheimer's accounts for 70 per cent of cases; vascular dementia, for 15 per cent; while the remaining 15 per cent includes frontotemporal dementia, Parkinson's, and a few rare types.

In fact, Alzheimer's can only be reliably established after death, following a pathological examination of the brain. People naturally want an earlier diagnosis so that the right treatment can be started. As a result, every form of dementia that cannot be put down to another cause (for example, where there is no evidence of Parkinson's or of small brain haemorrhages) is considered for convenience' sake as Alzheimer's. Generally, that will be a correct diagnosis, for the simple reason that Alzheimer's is the most common form of dementia.

In screening for dementia, doctors often administer the mini–mental state examination (MMSE), which samples functions such as memory, orientation to time and place, concentration, and language skills. A maximum of 30 points can be scored on the test, which is easily attainable for most healthy adults. A score of 27–30 is seen as normal, indicating no cognitive disorder. A score of 21–26 may be evidence of MCI or early-stage dementia. Of course, the reason for a lower score may be something completely different, such as depression, which also affects cognitive function. A score of 11–20 indicates moderate dementia, and scores below 10 indicate serious dementia. The MMSE is a short, general test, and in many cases, especially where the patient is suffering from mild impairment, a more extensive neuropsychological examination is advisable.

Short-term memory is the first casualty of Alzheimer's. If neuropsychological testing is done, it shows that verbal memory (such as recalling lists of words) is the first part of short-term memory to go. Memory problems are followed by a decline in some executive functions (the person may have difficulty planning activities and switching focus) and language problems (he or she may speak in a confused manner). Patients can also suffer from delusions or hallucinations. A delusion is when a person believes something that does not correspond to reality; a hallucination involves seeing or hearing something that isn't real. An example of a common delusion experienced by Alzheimer's patients is that a relative has stolen something from them. A hallucinating patient can see people, shapes, or objects that do not exist. In my book *Figments of Our Imagination*, I go into greater detail about changes in brain functions that can lead to hallucinations and delusions, particularly in psychiatric patients. I also show that they can occur, in a milder form, in healthy people. The parts of the brain involved in such experiences are the prefrontal cortex and the temporal cortex, precisely the areas most severely affected by Alzheimer's.

Vascular-dementia patients often have a history of (often tiny) brain haemorrhages or bleeding. There is an irregular pattern of cognitive decline that differs from patient to patient. One may primarily suffer from reduced cognitive flexibility or find it hard to plan and co-ordinate activities, another has the greatest difficulty in finding words, while yet another complains mostly of memory problems. This is related to the location of the haemorrhages.

Frontotemporal dementia is signalled early on by behavioural changes, including disinhibition. The part of the brain that is affected is not the hippocampus, as in Alzheimer's, but the prefrontal cortex, which regulates behaviour and suppresses inappropriate behaviour. Frontotemporal dementia is much less common than Alzheimer's but tends to occur at a younger age: if a person shows signs of dementia before their 65th birthday, it is more likely to be frontotemporal dementia.

The early stages of Parkinson's disease are often characterised by motor symptoms, such as trembling hands or difficulty getting out of a chair. Cells deep in the brain that produce dopamine begin to die. Dopamine is a neurotransmitter that influences mood, and the destruction of these cells explains the depression experienced by many Parkinson's patients.

Alzheimer's disease can be seen as an extreme form of the ageing process. Damaged neurons and accumulations of tangles and plaques also occur in normal ageing, but in Alzheimer's the damage is on a much greater scale. Cognitive functions are broken down one by one; memory problems are followed by thinking and language difficulties; and finally patients can no longer speak, walk, or make contact with others, and they become incontinent. This process takes years. Sometimes, as in the case of Ronald Reagan, it can take decades.

Most Alzheimer's patients at some point end up in a nursing home because their relatives can no longer care for them. An important question is what effect this has on their brains. Unfortunately, little research has been done into this issue. Although placing them in a home may often be the best thing for patients, it seems highly likely that this is not good for their brains. Admission to residential care is often accompanied by a reduction in physical exercise and mental stimulation. That needn't be the case: many nursing homes are now providing more activities for their residents, but the problem is often lack of staff. Family and friends could play a greater role in involving patients in stimulating activities.

The treatment for Alzheimer's consists of medication to combat cognitive decline, though the effect is only moderate. At the same time, behavioural therapy can help alleviate restlessness, sleep problems, anxiety, and aggression. The medication for Alzheimer's was described earlier in this chapter: drugs that reinforce the neurotransmitter acetylcholine or block the activity of glutamate. However, Professor Peter De Deyn, director of the Alzheimer's Centre at the University of Groningen since 2011, points out that this is merely symptom relief. Researchers worldwide are engaged in discovering the roots of the disease and finding ways to combat it, focusing on anti-inflammatory agents and medication that might reduce or prevent the protein build-up. Some success has been achieved in animal trials, but there has been no breakthrough where human subjects are concerned.

The breakdown of cognitive function in dementia is appalling to witness. Nevertheless, researchers at Alzheimer's centres in Nijmegen and Maastricht have recently urged us to see dementia in a more positive light.[10] They say that Alzheimer's is nothing more than an accumulation of phenomena associated with ageing that occur in healthy older people too, though to a lesser extent. Alzheimer's patients are above all our fellow human beings, rather than 'sick people'. In *The Old King in his Exile*, written by the Austrian author Arno Geiger about his father, he describes how life with Alzheimer's can still be meaningful and valuable, for both patient and relative.

To summarise: doctors and psychologists are now able to diagnose what is possibly a precursor of Alzheimer's disease. This early stage involves memory problems confirmed by someone close to the person in question. In 50 per cent of cases, the person does not go on to develop Alzheimer's. In those who do develop it, the hippocampus, an area of the brain vital to memory, is affected. Cognitive training and physical exercise can help slow decline. A major factor in Alzheimer's is the build-up of protein deposits in the brain. There is as yet no effective treatment, but medication and support can delay the deterioration to some extent.[11]

- Doctors' ability to diagnose MCI, which is in some cases a precursor of Alzheimer's, is constantly improving.
- In almost half of MCI cases, these problems remain stable. In one in seven cases, they actually improve so much that a diagnosis of MCI no longer applies.
- Cognitive training and physical exercise can help slow the decline associated with MCI.
- The hippocampus, an organ of the brain that is vital to memory, is affected in people who develop dementia. Alzheimer's is characterised by a large number of protein deposits in the brain.
- Age is the most important risk factor for dementia. One per cent of 60-year-olds, 7 per cent of 75-year-olds, and 30 per cent of 85-year-olds are sufferers.
- As yet, there is no effective treatment for dementia, but medication and support can slow the deterioration to some extent.

5

Body and Mind:

the influence of hormones

We are all aware that older people acquire wrinkles and have more fragile bones. We know that men in their seventies have weaker muscles than 40-year-olds. But do we know why that is? In fact, it's all down to hormones.

The word 'hormone' comes from the Greek *horm*, meaning 'impetus'. Hormones are chemical substances produced by the body and transported by the blood-stream to tissues and organs, stimulating or slowing their activity. One example is insulin, which regulates the concentration of blood sugar. It ensures that cells in the body take up glucose, reducing the levels in the blood.

As we grow older, the body's production and delivery of many major hormones diminishes. The most important of these are growth hormone and the sex hormones: oestrogens in women and testosterone in men. In women, the decline in production of oestrogens is called menopause, starting on average at the age of 50 and lasting about five years. A well-known symptom is what are called 'hot flashes'. These sudden feelings of heat in the upper body can be so overwhelming that sufferers will open all the windows in the middle of winter. A British actress about to be interviewed by *The Daily Telegraph* in a hotel lobby asked for the interview to take place outside on a bench in a bitter autumn wind — because of her hot flashes.

Other common symptoms of menopause include night sweats, bladder problems, dry skin, irritability, and fatigue. Expectations can also aggravate symptoms. Research has shown that women who have a negative attitude to menopause earlier in life often experience more symptoms when it actually arrives. Women in Eastern cultures, where it is less widely regarded as a problematic period, report fewer unpleasant symptoms than women in the West.

What is less well known is that hormonal changes in older men are also noticeable: this is sometimes called andropause or male menopause. Symptoms experienced by both men and women include reduced energy levels, mood swings, and diminished libido. Andropause is less severe than menopause. This is understandable: in women the whole hormonal reproductive system gradually shuts down, while in men there is only a slow decline in production. Incidentally, as with many other signs of ageing, not everyone experiences problems.

HORMONAL CHANGES

We don't often associate sex hormones with our brains. Nevertheless, declining hormone levels have an unfavourable effect on our mental abilities. When Pam, a businesswoman in her mid-fifties, entered menopause, she worried for a time that she was suffering from early-onset dementia. Sometimes she couldn't even think of a simple word such as 'cat'. She had just started a course on mortgages (she was a successful financial adviser to banks) but noticed that she wasn't absorbing or retaining any of the information. When she signed up for the course, she thought it would be a cakewalk, but in the end she didn't take the final exam.

A study conducted at the University of California of over 2000 women showed that Pam is not alone. Memory problems increase early in menopause. Many women process information more slowly in this period. Mostly, this is not an alarming decline — they can still find their way around town, for instance — but they notice that their thought processes are fuzzier than before. The good news is that a year after their last menstruation (when menopause is over), memory, concentration, and the ability to learn will improve to some extent and will be less of a problem than during menopause. But first, let's take a look at how these hormones influence the brain.

OESTROGENS AND MENOPAUSAL SYMPTOMS

Oestrogens are known as the 'female hormones'. In fact, men also produce these hormones, but oestrogen levels are much higher in women. The production cycle of oestrogens and many other hormones resembles the path of a boomerang: it begins in the brain, and eventually returns to the brain. The hypothalamus begins the process. The signal to start secreting oestrogens passes a couple of intervening stations, including the pituitary gland, before arriving in the ovaries (figure 16).

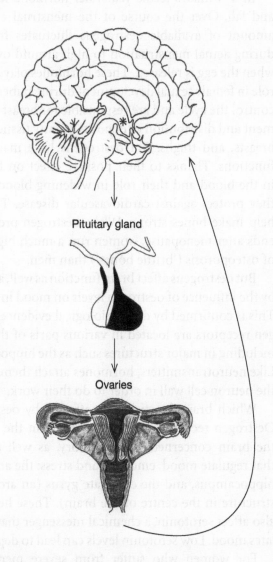

Pituitary gland

Ovaries

FIGURE 16. *After receiving a signal from the brain (hypothalamus), the ovaries begin to produce oestrogens.*

In a woman's fertile years, her hormone levels rise and fall. Over the course of the menstrual cycle, the amount of available oestrogens fluctuates from low during actual menstruation to high around ovulation, when the egg is released. These hormones play a crucial role in female sexual development during puberty. They control the cycle and are responsible for breast enlargement and distribution of deposits of fatty tissue in hips, breasts, and thighs. But oestrogens have many more functions. Thanks to their positive effect on fat levels in the blood and their role in widening blood vessels, they protect against cardiovascular disease. They also help make bones strong. When oestrogen production ends after menopause, women run a much higher risk of osteoporosis ('brittle bones') than men.

But oestrogens affect brain function as well, as shown by the influence of oestrogen levels on mood in women. This is confirmed by direct biological evidence: oestrogen receptors are located in various parts of the brain, including in major structures such as the hippocampus. Like neurotransmitters, hormones attach themselves to the neuron cell wall in order to do their work.

Which brain functions are affected by oestrogens? Oestrogen receptors are mainly found in the areas of the brain concerned with memory, as well as those that regulate mood, emotion, and stress: the amygdala, hippocampus, and the cingulate gyrus (an arc-shaped structure in the centre of the brain). These hormones also affect serotonin, a chemical messenger that modulates mood. Low serotonin levels can lead to depression.

For women who suffer from severe menopausal symptoms, prescribing oestrogens would seem an obvious answer. And this was in fact the practice for many years. Research has shown that it had a positive effect in reducing hot flashes and improving sleep, mood, and general wellbeing. Even memory improved in some cases. Studies using brain scans established that prescribing oestrogens after menopause helps activate the prefrontal lobe, improving working memory. Unfortunately, this type of hormone therapy has serious side effects: in some women, the risk of developing certain forms of cancer (ovarian cancer, for instance) is increased. Oestrogens promote the growth of tumour cells. What's more, the therapy increases the risk of cardiovascular disease.

For all these reasons, doctors and scientists have been searching for other substances with the positive effects of oestrogens but not the negative. Phytoestrogens are a natural candidate, both literally and figuratively speaking. These compounds, obtained from plants, have the same structure as oestrogens and would appear to have the same effects. However, they do not promote the growth of tumour cells and have no significant side effects.

The soybean plant contains abundant quantities of phytoestrogens. So does eating lots of soy relieve meno-pausal symptoms? This would seem to be the case. In Asian countries, for example, where soy is an import-ant part of the diet, women usually have fewer symp-toms, both mental and physical. Though that is a very interesting fact, it is doesn't necessarily mean that this is the result of eating soy. Other, cultural differences may be implicated. Yet the assumed link with soy is not unreasonable: animal research has demonstrated that the oestrogens in soy can have biologically comparable effects to those of mammalian oestrogens.

A number of studies have shown that eating soy can slightly reduce hot flashes in women who suffer severely from them. The best way to find out if soy products can also have a beneficial effect on memory and concentration would be to take a group of women in menopause who eat soy daily and compare them with a control group receiving a placebo. After a certain amount of time has elapsed, researchers see whether the women in the first group indeed suffer fewer symptoms. This is what a team led by Professor Yvonne van der Schouw at the University Medical Centre Utrecht did. Together with Professor Edward de Haan, I was responsible for the choice of neuropsychological tests and interpretation of the scores for this study. The subjects were 202 women in menopause, assigned on a random basis to the soy group or the placebo group. In the first group, the women consumed a soy preparation, consisting of a powder that could be added to drinks or to food, every day for a year. The placebo also consisted of a powder, this time containing milk proteins but no active ingredient.[1]

Neuropsychological tests were administered before and after the trial year, assessing attention, memory, reasoning, and the ability to plan and switch focus. Other tests measured bone density and blood fats (such as cholesterol), areas where oestrogens also have a beneficial effect. The researchers eagerly awaited the results of all these analyses: would the phytoestrogens in the soy preparation improve cognitive function? Disappointingly, the answer was no. In neither group was any improvement measured, either in cognitive function or bone density and fat levels. In 2004, the results of the study were reported in *JAMA*, the authoritative journal of the American Medical Association. Later research, including a Chinese study of 191 women who took either soy or a placebo for six months, also failed to demonstrate any improvement in memory and other cognitive functions. Recently, the absence of any effect from phytoestrogens was confirmed once again, in a study conducted at the University of Miami in which nearly 250 women took part. One group took pills on a daily basis containing twice the amount of soy normally consumed by people who eat a soy-rich diet. The other group took a placebo. The researchers found no improvement in bone density, quality of sleep, or the frequency of hot flashes. Unfortunately, they did not measure cognitive function.

Thus, phytoestrogens have not lived up to their early promise. I would add, though, that the most recent studies also question the effectiveness of regular oestrogen therapy. This, too, does not seem to bring about a long-term improvement in mental capacities, perhaps because hormone levels are being raised artificially. The oestrogens produced by the body before menopause are released into the bloodstream in a different manner, spread more evenly over the day, but with peaks and troughs. It is also possible that the number and function of oestrogen receptors declines after menopause as a result of diminished oestrogen production, and that this is the reason oestrogen therapy has less effect.

TESTOSTERONE AND COGNITIVE FUNCTIONS

Testosterone quite often gets a bad press. Almost everyone knows that it is a male hormone, abused by some athletes, particularly bodybuilders, and associated with aggression and annoying behaviour. If teenage boys are loud and insolent or get into fights, we put it down to excess testosterone. The same goes for macho behaviour.

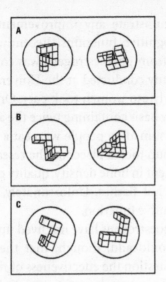

FIGURE 17. *Test of spatial insight: is the figure on the right the same as the one on the left?*
Answers: A, Yes; B, Yes; C, No

It's true that testosterone is the most important male sex hormone. It builds muscle and may in some cases have an influence on whether someone actually makes use of those muscles. But that's not the whole story. Testosterone is also important for skin and bones, a healthy sense of competition, the male sex drive, and the production of sperm cells. It's a very 'physical' hormone. Like oestrogen, it illustrates the close relationship between mind and body. Testosterone influences mental capacities such as memory and concentration. If you measure the testosterone in the blood of 100 men and then ask them to take a number of neuropsychological tests, you'll see a correlation between the amount of testosterone and performance: men with more testosterone score better, mostly where spatial insight is concerned. One way of testing spatial insight is to ask people to decide whether two abstract figures drawn from different perspectives are the same (figure 17).

To answer the question, people rotate the figure in their minds so they can see if it is the same as the figure in the second drawing. On average, men are better at this than women, while men with a lot of testosterone score better than men with less.

Just as men produce little oestrogen, women produce little testosterone. To test the influence of testosterone on cognitive functions, a single testosterone pill was given to young women in a number of studies to see if their performance changed. If women were to take such a pill every day for a long period of time, they would develop male characteristics such as increased facial hair and deepening of the voice. But a single pill is harmless. In a study we conducted with psychologist Dr Jack van Honk, a specialist in the influence of hormones on human behaviour, we examined the effect of a single testosterone pill on spatial insight. This improved in the female participants after they ingested the pill, compared with those who took a placebo. This shows that testosterone can have a powerful influence on cognitive functions, something that other studies have also demonstrated. In the laboratory, we used to joke that it might be handy for our wives and girlfriends to take the testosterone pill before reading maps in the car while on holiday.

From about the age 50, testosterone production in men starts to decline: andropause begins. Although hormone levels diminish less radically than they do in women, this can still have serious effects, ranging from a loss of muscle mass and strength, and reduced sex drive (erection problems, for instance), to sleeplessness, fatigue, and depression. But is this accompanied by a decline in cognitive functions? Dr Majon Muller researched this question at the University Medical Centre Utrecht in a study involving 400 men between the ages of 40 and 80 (100 subjects for each decade). I worked on the neuropsychological part of the study. A link was established between testosterone levels in the blood (without any testosterone being administered), and memory performance and speed of information processing.

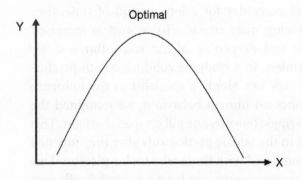

FIGURE 18. *Inverted U-shaped curve reflecting the relationship between testosterone levels and cognitive functions in men aged between 40 and 70. Men with very low and very high levels perform less well in neuropsychological testing than men who have intermediate levels. X-axis: testosterone levels; y-axis: scores in neuropsychological testing.*

What was remarkable about this result was the fact that the relationship was *curvilinear*: it had the shape of an inverted U (see figure 18). A curvilinear relationship indicates that there is an optimal level (in the middle of the inverted U). A linear relation (that is, the higher the better) was only found in the oldest group, the men between 70 and 80.

For men in this age group, whose testosterone levels are generally low, the rule of thumb is that the more testosterone they have, the better their memory and speed of processing. This raises the question of whether giving them testosterone would lead to improved cognitive functions, a hypothesis that was also investigated by researchers at the same institute, with whom I worked.[2] Again, we used a range of neuropsychological tests to map memory performance, processing speed, and cognitive flexibility. Because the study centred on testosterone, we added an extra test of spatial insight using the same objects as in figure 17. The participants were a group of 236 men between the ages of 60 and 80 with low testosterone levels. This was a requirement for participation, since it provided a biological reason for supplementing the hormone. For six months, the participants took a twice-daily pill containing testosterone or a placebo. There was no greater improvement in cognitive skills in the group that took the testosterone compared with the group that took the placebo. The men in the testosterone group displayed a decrease in fatty tissue but no increase in muscle strength.

So it seems that increasing low testosterone levels in older men does not automatically result in improved memory or concentration. We found no improvement even in spatial insight, which is more closely linked to testosterone levels in younger people than memory and concentration are. It is of course possible that six months is too short a period for changes in cognitive functions to occur, though many substances and hormones produce changes within a few weeks. Ingesting pills results in temporary increases in hormone levels that persist for several hours afterwards. This is different from natural hormone levels, which may have an impact on the pills' effectiveness, or even result in a lack of effect. Testosterone plasters and gels that can produce a more stable increase in hormone levels have recently come onto the market. Though more research is needed, there are currently no indications that these have a positive influence on cognitive function.

Another question that deserves further study is whether testosterone might lead to improved cognitive skills in people who score low on neuropsychological tests. In our study, the participants scored just above average for their age while their testosterone levels were lower than average. This meant two things: first, that there was plenty of room for improvement; but second, that they had no greater memory and concentration problems than their peers. It is possible that people who do have these problems would benefit from hormone treatment. The aim of this type of treatment is to improve a clearly reduced cognitive ability, such as poor memory.

Commercial organisations go further and play on fears of growing older and a general decline in bodily functions. In line with the worship of youth in America (and increasingly in Europe), the American age-management physician Dr Jeffry Life advocates the use of hormones to ensure a well-muscled body in your seventies. He is himself the living proof of his theories. Though born in 1938, he boasts a tanned body, heavily muscled arms, a broad chest, and a six-pack. People who see the before-and-after photos he uses to advertise his business often think that the images have been photoshopped — that the head of a 70-year-old has been placed on the body of a 30-year-old Olympic athlete. But Dr Life is real; even his name is real. He transformed his body through working out at the gym, a special diet, and treatments with testosterone and growth hormone. You may wonder, of course, if it is a good thing to fight off the natural ageing process in this way. In any event, the usefulness of hormone therapy has not been scientifically proven, especially when it comes to brain function. And although it can help you develop a more highly muscled body, it can also have unwanted side effects. The use of testosterone may activate latent prostate cancer. And the risk of side effects is even greater when it comes to growth hormone, perhaps the most important hormone that declines as we grow older and a major contributor to the symptoms of ageing.

A MYSTERIOUS HORMONE

The 17th-century French philosopher René Descartes is famous for his statement 'Cogito ergo sum' (I think, therefore I am) and for his take on the concept of dualism, which holds that body and mind are two separate things. He argued that body and mind interact through the pineal gland, a small organ located near the centre of the brain between the two hemispheres. If Descartes were alive today, he would probably have regarded growth hormone as the link between mind and body. This is the most 'physical' hormone we produce, but it nevertheless has an enormous influence on our minds. Faced with our current knowledge of this hormone, Descartes might even have come to doubt the theory of dualism.

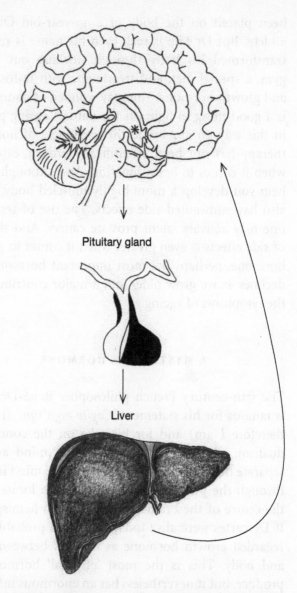

Pituitary gland

Liver

FIGURE 19. *The hypothalamus stimulates the release of growth hormone by the pituitary gland, which in turn signals to the liver to start producing insulin-like growth factor (IGF), which has a beneficial effect on the brain.*

As its name suggests, growth hormone is what makes the body grow from infancy into adulthood. People with too little remain small; people with too much can grow to well over six feet tall. But it continues to play a major role throughout our lives. It encourages the formation of fat-free tissue in the body, which is healthy. It helps to renew skin and hair, and to repair damaged tissue. Less well known is the fact that this natural hormone has a positive influence on brain function. People with high levels of growth hormone feel better than those with low levels: they literally and figuratively feel more comfortable in their own skin. We have only known for about 20 years that growth-hormone levels correlate with scores in memory, concentration, and thinking skills. Before we look at that more closely, let's look at how growth hormone actually works.

As in the case of sex hormones, the signal to start producing growth hormone is sent by the hypothalamus to the pituitary gland on the underside of the brain. Cells in the pituitary produce and release growth hormone in a series of irregular pulsations into the bloodstream. Release peaks about an hour after the onset of sleep, which is one of the reasons why this stage of sleep, characterised by deep sleep, is so important. The levels of growth hormone in the body decline as we age, halving every ten years after adulthood. Its effect on organs and tissues is mediated by an intriguing hormone related to insulin and therefore called 'insulin-like growth factor' (IGF), the last piece in the domino effect initiated by the hypothalamus (see figure 19). It is difficult to reliably measure growth-hormone levels in the blood because the pituitary releases it in the irregular pulsations described above. Levels of IGF, on the other hand, are much more stable. And because

IGF is the chief slave of growth hormone, obeying its orders to bind with cells in tissue, many studies simply measure IGF as an indication of the activity of growth hormone. The remarkable thing about IGF is that it works throughout the body, including the brain. This is unusual because most substances elsewhere in the body cannot cross what is called the blood–brain barrier. But IGF is involved in the growth of brain cells and the dendrites formed when we learn new information. It is also essential to the repair of damaged brain tissue.

I first heard of IGF in 1997 when I was doing my final project for my psychology degree. I was supervised by Dr Hans Koppeschaar, a specialist in internal medicine at the University Medical Centre Utrecht. His fascination with the possible influence of IGF on the brain was infectious, and I, too, became interested in this mysterious hormone. Both IGF and cognitive function decline with ageing, and our question was whether there was a connection between the two.[3] There were two reasons for investigating this only in men: first, because there may be differences between men and women in the way growth hormone works; and second, because we wanted to see if our findings were in line with the single previous study of IGF and cognitive function, an American study using older men.

The men in our study were between 65 and 76 years of age. The results showed a relationship between IGF levels and performance on two tests: one measuring processing speed; the other, cognitive flexibility. The more IGF a participant had in his blood, the better he scored on the tests. The earlier American study had also found a link between IGF and cognitive flexibility. Later research, using much larger groups, has confirmed that higher IGF values in healthy older people are associated with better cognitive performance. Researchers in Amsterdam, for example, found that people with the lowest IGF values in a group of over 1,300 participants (aged between 65 and 88) had the lowest scores on speed-of-processing tests. This was true of both men and women. The low IGF values also proved to be linked to reduced processing speed three years later.

The fact that growth hormone and IGF affect cognitive abilities did not come as a complete surprise. It had already been discovered that people with extremely low levels of growth hormone perform less well on neuropsychological tests, particularly those measuring memory and concentration. And it has recently been established that people who produce extremely high levels of growth hormone (a disease known as acromegaly) suffer from memory problems. As is the case with testosterone, both too little and too much of this hormone appears to have a negative effect on the brain.

It might seem logical to administer extra hormones to people with low levels to increase those levels, but there are risks involved. Extra growth hormone encourages the growth of any cancer cells that might be present but are as yet undetected. And that risk is higher for older people. That is why little research has been done in this area. In fact, only three studies have been published: one in which growth hormone was given to older people with low hormone levels, another where IGF was given to a group of older women, and a third in which the substance released by the pituitary gland to stimulate growth hormone secretion (known as growth-hormone-releasing factor or GHRF) was administered. The first two studies showed no improvement in cognitive function, though the study using GHRF did. Growth hormone works by setting off a chain reaction in the body, and intervening as early as possible in the chain (GHRF comes in at the start of the chain reaction) might achieve the best results.

Caution with regard to administering growth hormone to older people is justified. In my view, further research should focus on people with memory problems or mild cognitive decline because their needs are more compelling: to halt the progress of Alzheimer's at an early stage.

IMPORTANT INSIGHTS
- Hormones play a crucial role as we grow older. A decline in hormone levels leads to a whole range of signs and symptoms of ageing.
- Hormones can be used to treat these symptoms, but to date there has been little success in improving cognitive function.
- Phytoestrogens in soy products do not help to combat problems associated with ageing.
- Growth hormones could have a positive effect on memory and concentration, but also present health risks.

6

Pills, Powders, and Push-ups:

what works (and what doesn't)

Gerrit Deems from Nijmegen didn't regard it as much of an achievement himself, but of course it was. On 19 April 2011, he obtained his PhD at the ripe old age of 89. Having worked as a sales representative and later as a social worker, he began studying philosophy when he was 43 and completed a theology degree at the age of 69. He then decided to expand his final paper, about the priest Alfons Ariëns (1860–1928), a key figure in the Catholic trade-union movement in the Netherlands, into a thesis. In an interview in the newspaper *NRC Handelsblad*, he admitted he had had his 'ups and downs' while writing the thesis.[1] But the dozen or so young PhD students I have supervised in recent years had the same problem. Deems's thesis, of course, gave him the perfect opportunity to keep his mind sharp. If people don't stay active as they age, they can stagnate. And he agrees: 'Being retired actually creates the space to do your own thing and to develop your talents even further.'

Advising everyone over the age of 65 to write a thesis is possibly going a step too far. But there are many other activities that make demands on your brain. So the question is this: are intellectual pursuits good for the ageing brain, and do they prevent decline? The Roman philosopher and statesman Marcus Tullius Cicero (106–43 BCE) wrote: 'I have no doubt that it is [impaired] in persons who do not exercise their memory.' Was he right?

All kinds of remedies are advertised on TV, on the internet, and in newspapers and magazines to keep your brain sharp as you age — from Sudoku puzzles and computer games to omega-3 fatty acids and extra vitamins. Scientists and pharmaceutical companies are busy developing medications to improve memory and other cognitive abilities. But do these things work? Is it in fact possible to improve cognitive function, or can we at best achieve a delay in its decline? In this chapter, we'll take a look at what works and what doesn't.

PILLS

British professor Linda Partridge is a founding director of the Max Planck Institute for Biology of Ageing in Cologne, and plays a pioneering role in research into ageing. While in the Netherlands for a lecture in November 2011, she gave an interview to *de Volkskrant*. According to Partridge, advances are being made in research into healthy ageing, and in a few years a special pill for daily consumption by older people will come onto the market. This pill will ensure that they remain healthy for longer. She called it a 'polypill' because it will have multiple ingredients, including an anti-cholesterol drug and aspirin (both reduce the risk of cardiovascular disease). It will probably also contain something that keeps insulin levels stable, which is good for metabolism. Many pills for older people already exist, but they produce only a single effect, such as the anti-cholesterol drug, and then only in a certain group of people (those whose cholesterol levels are too high, for example).

What Professor Partridge did not mention were medications that would directly improve cognitive skills. That is because these are not yet sufficiently developed. It is even unclear whether it will be possible to improve, say, memory and concentration with a single pill without side effects. But lots of research is going on into substances known as cognition enhancers, which can improve memory and other cognitive abilities. Examples include donepezil, memantine, and modafinil. The first two of these are already being used in the treatment of Alzheimer's. Small improvements in memory and concentration have been noted in patients taking these drugs, but though they sometimes slow the rate of decline a little, they do not stop the progress of the disease.

Donepezil increases the amount of acetylcholine (which is important for memory) available in the brain. As yet, little research has been done into its effectiveness in healthy seniors with memory problems. An American study of 20 people aged 72 showed that it improved the ability to recall words, but only if the person had thought about the word first. In this experiment, people were asked to say if they thought a particular word was pleasant or unpleasant. It is easier to recall words you have 'processed' in this way than if you merely read them. Donepezil reinforces this mechanism, improving storage in the brain. Another study of people aged between 60 and 80 established that this drug promoted REM sleep (when most dreaming takes place) and memory consolidation. The two proved to be linked. This is very interesting because an increasing number of studies have shown that, even in younger people, REM sleep is important for retaining relevant information and experiences. Another study, which used a driving simulator, investigated whether donepezil could help older people remain alert while driving. However, the drug proved no more effective than a placebo.

Memantine affects three neurotransmitters simultaneously: glutamate, serotonin, and acetylcholine. Glutamate is widely present in the brain and activates neurons. Like acetylcholine, it is vital to learning and memory. Though there is sufficient evidence from animal research and studies of Alzheimer's patients that memantine can have a beneficial effect on cognitive function, the drug has not yet been properly tested in healthy people with memory problems. Currently, a new class of compounds called ampakines is being tested on animals in experimental research. Ampakines also act via glutamate receptors and appear to facilitate learning and memory.

Finally, there's modafinil, which is already used as a medication for people with sleep disturbance (narcolepsy). Because patients became more alert after taking the drug, researchers decided to investigate whether it could improve concentration in people suffering from cognitive decline. We don't exactly know how modafinil works in the brain, but there are indications that it increases levels of the neurotransmitters dopamine and histamine. Histamine makes us alert, and dopamine improves attention and working memory. Research has shown that modafinil can enhance both these functions in young, healthy people who take it once as part of a study, as well as in people with ADHD or schizophrenia. No research has as yet been done using healthy seniors with memory problems or people with MCI.

To conclude this section, we should look at another pill that can improve memory: the placebo. Researchers at Victoria University of Wellington in New Zealand gave one half of a group of older people a pill they were told was good for the memory. In reality, it was a sham drug, with no active ingredient. Next, they were set a demanding memory task, which involved recalling a series of actions in the right order. Subjects who had taken the pill performed better. This study confirms that expectations play an important role in cognitive performance. We saw earlier that stereotypical ideas about older people and memory loss in ageing are also common among older people themselves, and these can affect their performance. Though placebo research shows that sham pills can reduce negative expectations, improving awareness in society might be more effective in the long run. Up to now, the positive aspects of ageing have received little publicity, but more knowledge about the ageing brain could lead to a reassessment and to fewer stereotypes. I hope this book will contribute to that process.

NATURAL SUPPLEMENTS

People have always tried to improve their health by using medicinal herbs. There is already a vast storehouse of information on their use in China, from thousands of years ago. Other cultures have also attached great value to herbs. For example, the *Wisdom of Sirach*, a work of ethical teachings from the 2nd century BCE, says: 'The Lord has created medicines from the earth, and a sensible man will not despise them.' And indeed, contemporary Western society is far from despising them. Many people buy herbal preparations from pharmacies and health-food shops, without always being aware that conventional medicines often contain natural substances with medicinal effects. Herbal supplements can be effective in treating physical complaints and psychological problems — St John's wort, for instance, can be useful in combating mild to moderate depression. But can they improve memory, concentration, and thinking skills?

Though claims are made for all kinds of products that supposedly act on the brain, only two have been scientifically proven to improve function or delay decline: vitamin B12 and omega-3 fatty acids. There is much evidence that B12 is important to cognitive function in older people.[2] Vitamin B12 deficiency is associated with a greater risk of cognitive decline in ageing, and around 15 per cent of older people have such a deficiency. Symptoms of low B12 levels include pallor, a feeling of weakness and fatigue, light-headedness, and a lack of appetite. Of course, if you notice these symptoms, they can be the result of a different condition, so it's important to consult your GP. A deficiency is unlikely in people who eat a varied and healthy diet. Blood tests are needed to establish B12 levels, and if there is a serious deficiency, your doctor may prescribe supplementation in the form of injections. A recent study conducted by the University of Chicago showed that a B12 deficiency in older people is associated with less grey matter in the brain. This finding confirms how important it is to consume foods rich in B12 (for example, beef or cod). B12-enriched soy milk is also readily available, while a yeast extract such as Marmite or Vegemite contains added B12, too.

Yet there is no hard evidence that taking extra B12 improves cognitive function. Not enough good research has been done in this area. Researchers in Australia asked 900 people between the ages of 60 and 74 to take either folic acid and B12, or a placebo every day for two years. The group that took the B12 and folic acid did better in memory tests than the group who took the placebo. In fact, folic acid alone can lead to improved function. A Dutch study involving 818 subjects who took folic acid or a placebo on a daily basis for three years found that the group taking the folic acid scored better on memory and processing speed. But there have also been studies that found no improvement in people taking B12. In French research involving over 800 people between the ages of 45 and 80, half of the subjects took a combination of vitamin B12 and omega-3 fatty acids; the rest, a placebo. All of the participants had cardiovascular problems, varying from heart attacks to brain haemorrhages. After four years of daily ingestion, neuropsychological testing found no differences between the two groups. Only the people who had experienced a brain haemorrhage improved their performance, and then only on a single test. The B12 taken in the Australian study might have had more effect because it was combined with folic acid, which promotes B12 uptake. Researchers at Wageningen University have established that folic acid and B12 can be added to bread. If all older people were to eat this enriched bread, B12 deficiency in this group could be eradicated completely.

Taking B12 to improve mental capacities may only be effective in people who have a deficiency, rather than in the general population. Though strictly speaking they are not suffering from a deficiency, a quarter of seniors have low levels of this vitamin, so supplements are advisable. Your GP or local pharmacist can advise on whether extra vitamins are necessary, and if so, in what amounts.

Omega-3 fatty acids have already been mentioned. They are found in oily fish such as mackerel, herring, and salmon, and are good for the heart and blood vessels. They also have a beneficial effect on the cell wall in neurons, enabling better transport of substances needed by the neuron to function effectively. A study among 107 people (average age 78) showed that those who regularly ate oily fish performed better in memory and concentration tests and had more grey matter in their brains. The improvement was linked to the levels of omega-3 fatty acids measured in the blood. Another study followed a group of older people over a period of 13 years, recording their consumption of fish. People who ate more fish complained less of cognitive decline. There may thus be a link between fish oil and improved cognitive function, but, once again, the key question is this: can you improve your mental capacities and prevent Alzheimer's by regularly eating oily fish or taking daily fish-oil capsules? Research using apes and rats found that memory did improve, which means that there is, at the very least, a possibility of benefit to the brain. But research on humans has yet to give an unambiguous answer.

A couple of studies did find an improvement in human memory. One, a Chinese study using almost 1,500 subjects over the age of 55, found a link between consumption of fish-oil capsules and fewer cognitive deficits after 18 months. But there are also a number of high-quality studies that reported no improvement, even after 26 weeks of fish-oil consumption. So though there is no guarantee that it will improve our memories and concentration, eating oily fish regularly (the guidelines say twice a week) appears to be good for your brain. And there is evidence that it is good for your cardiovascular system.

Another natural supplement claimed to be beneficial for the ageing brain comes from *Ginkgo biloba*, a tree species originating in China. The tree can live as long as 1,000 years and has distinctive fan-shaped leaves, from which ginkgo is extracted.[3] The herb helps the body to rid itself of free radicals, waste products that can damage brain tissue. Ginkgo is thus an antioxidant. It can also protect the structure of cell walls in neurons, and it promotes healing in damaged areas of the brain by stimulating the uptake of glucose. Furthermore, an Australian study showed that ginkgo can strengthen brain-wave activity related to working memory, and the researchers thought that this was related to a more efficient working memory.

Though these findings primarily come from animal research, it is not unlikely that ginkgo has similar beneficial effects in humans. A couple of studies have found small improvements in the memory of Alzheimer's patients, but a similar number found no improvement. The number of participants in these studies was too small for reliable conclusions to be drawn. So far, too little research has been done among healthy seniors.

Other substances that can temporarily improve concentration include coffee and ginseng. Like ginkgo, ginseng (*Panax ginseng*) comes from China. This time, the active ingredient is in the roots rather than the leaves of the plant. Even though it is one of the most commonly used herbs in the world, very little high-quality research has been done into its effects in humans. This is possibly because there is little to be earned from ginseng, a natural product that cannot be patented — the pharmaceutical industry prefers to invest in large-scale research into new medicines because they can recoup the investment through sales. The active ingredients of ginseng are ginsenosides (also known as panaxosides). Animal research shows that these can have a positive effect on acetylcholine levels in the brain. They can also activate energy metabolism in the hippocampus.

COGNITIVE TRAINING

An advertisement for the popular Nintendo game *Dr Kawashima's Brain Training* features a fit-looking 50-year-old who smiles as the screen reports that he has a 'brain age' of 46. Just before receiving this good news, he has solved several mathematical and other types of puzzles. The computer keeps track of his score and compares it with those of other players. In another advertisement, you see a woman of about 50 looking disappointed when she learns that her mental age is 63. But there's hope: she can improve her mental abilities with *Brain Training*!

The idea of 100-year-olds concentrating all evening on their Nintendo games is perhaps rather strange. But that is exactly what British centenarian Kathleen Connell does, according to *The Telegraph*. 'It's absolutely super, I don't know what I'd do without my Nintendo,' said Connell, who plays other games on it, including Scrabble, to keep her mind fit. *Brain Training* consists of different tasks to train memory, concentration, and speed of processing. Some take the form of sums, puzzles, or language lessons, while others are more like standard computer games. In one of the tasks, you have to count how many people are on the screen, and then a house is placed over them. Some people leave the house and others enter, and the player has to keep track of the number of people in the house. Though *Brain Training* was developed in collaboration with a scientist (you guessed it — Dr Kawashima), there is to date little scientific evidence that it improves older people's cognitive functions or prevents their decline. This is surprising for a game designed to do exactly that, and which has sold like hot cakes. Nintendo, however, was wise enough never to claim that it could prevent cognitive decline in older people. It certainly can do no harm to play games such as these on a regular basis, but I do object to the calculation of a mental age — this is just not possible on the basis of a few short tests on a games console. And it is irresponsible to tell someone who is 50 that they have a mental age of 63 without any explanation from an expert of what that means.

Is there any cognitive training that has been properly researched? In fact, cognitive training is nothing more than practising mental capacities such as memory, concentration, and thinking skills. People who have worked their whole lives with their brains because their profession required a lot of reading and reflection have a lower risk of developing Alzheimer's. That in itself points to the positive effect of mental exercise in the long term.

The results of short-term exercises for older people have been studied in recent years. These exercises mainly consist of memory training, but differ between studies as each researcher invents their own variant. This makes it difficult to come to any general conclusions. Nevertheless, certain trends can be discerned. Researchers at the University of Southern California analysed the results of all published research into cognitive training for older people. It emerged that training can improve cognitive functions by 10 per cent. Given that the brain's decline between 65 and 75 is on average 10 per cent, this is substantial. It doesn't mean, however, that cognitive training can fully compensate for that decline. Cognitive training is often extremely specific, while the decline is across the board. So if you do a lot of Sudoku puzzkes, you become extremely good at Sudoku puzzles, but your brain hasn't necessarily become sharper in other areas.

Nonetheless, there are indications that training based on lots of different exercises can improve cognitive skills generally. For example, a complex computer game can do this, as shown by American psychology professor Arthur Kramer and researcher Chandramallika Basak, both at the University of Illinois. They trained 20 older people in a real-time strategy game, *Rise of Nations*, and compared their performance in neuropsychological tests afterwards with that of 20 of their peers who had not been trained. In *Rise of Nations*, players have to run their own country: they build cities, help people to find work and food, maintain an army, and expand their territory. This makes a great demand on executive functions such as working memory, planning and calculating, maintaining oversight, and strategic thinking. In the end, the participants who had played the game (for an average of 23.5 hours) scored better on the tests than the control participants.

Though little research has been done in this area, physical changes in the brain after cognitive training have actually been measured. A German study found that nerve connections between the brain hemispheres were strengthened after six months of memory and speed-of-processing training in a group of people with an average age of 69. And researchers at Wake Forest University in North Carolina reported an increase in blood flow to the prefrontal cortex after eight weeks of cognitive training in a group of people aged around 70. They were compared with a group of the same age who attended interactive lectures on health and ageing. The increased blood flow to the prefrontal cortex after training was accompanied by improved concentration and ability to ignore distractions.

But cognitive training has an added benefit: older people actually *feel* better as a result. A Spanish study demonstrated that a group of seniors who had undergone cognitive training not only performed better in neuropsychological testing but also reported improved quality of life — a concept used in science to describe an individual's general wellbeing and contentment with their situation. This may be the result of the satisfaction that comes from understanding things slightly faster or being able to recall slightly more. Or you may feel that you have more control over your cognitive abilities because you've worked on them.

In general, memory training has led to small improvements, especially in the skill that was practised the most in these studies: the ability to recall words and stories. But good training doesn't just involve memorising. It also gives the participant useful strategies, such as the loci method mentioned earlier (in which you link names or items you have to remember with locations, for instance in and around the house). Another effective strategy is to focus more strongly on retrieving the information than on storing it. Imagine you want to memorise a piece of text for a speech. You can read it 20 times so you know it by heart, a method you'll remember from your schooldays. But it's more effective to read it three times and then try to recall what you can without consulting the written text. You may have to check a few times, but if you do, put the paper away again immediately afterwards. This is a more active way of learning that helps you to see a structure in the text, which is what aids memory. Research conducted at the Johns Hopkins Bloomberg School of Public Health in Baltimore among 1,401 older people who received memory training showed that they learned to make better use of memory strategies, thereby improving their memories. Follow-up research established that these improvements were maintained over a period of five years. Furthermore, the training had a positive effect on everyday functioning, including living independently, finding their way around, and remembering crucial things.

There are even more ways of enhancing learning ability — for instance, making crazy connections in order to remember things. Imagine you have to take your grandson to football practice on Beethoven Street. To remember that, you could picture Beethoven sitting at the piano, playing a sonata with a football balanced on his head while your grandson looks on in amazement. A good mood also helps with learning, as a study conducted by Professor Richard Ridderinkhof of the University of Amsterdam has shown. If you feel good, you absorb information better. But how do you achieve that? One way is to think about happy or amusing events in your life. Putting on your favourite music can help, too; or promising yourself a (healthy!) reward once you've completed the task.

How long do the effects of cognitive training last? Various studies have shown that this can be up to two years in the case of memory training, provided it was intensive: half an hour three times a week for at least three months. It is also advisable once training ends to continue to challenge yourself every day to remember things such as the names of people you meet, your shopping list, or what you had for dinner a couple of days ago. We don't know if such exercises can (at least partially) slow the general decline in memory. More research is needed.

It is unlikely that training will completely halt cognitive decline in the long term, but the fact that your memory functions better if you exercise it is a valuable insight. Nor do we know if only the skill you actually exercise improves, or whether there is a more general effect. For example, you've learned a great number of word lists and improved your verbal memory. Does that mean you've only improved your ability to learn lists, or can you now remember where the soft drinks are in the supermarket? After all, the latter is a question of spatial memory, which you haven't trained. To date, it would appear that exercises mainly improve the skill that has been trained.

But there are also studies demonstrating that older people can achieve better results on tests they have not been trained in by practising other, comparable skills. This seems mainly to apply to training that involves learning skills that are multifaceted and complex. Examples include learning a new language or a musical instrument. Making music is very good for cognitive function in both children and adults. A study of 70 healthy people between the ages of 60 and 83 with varying degrees of musical knowledge showed that those who regularly played an instrument scored better on a variety of neuropsychological tests. The longer a person had played (many had started when they were around ten years old), the better the scores. According to researcher Brenda Hanna-Pladdy of Emory University, 'Since studying an instrument requires years of practice and learning, it may create alternate connections in the brain that could compensate for cognitive declines as we get older.'

Another suggestion for realistic training comes from neuropsychologists at Washington University, St Louis, who propose a course in birdwatching. There are two benefits here: first, your memory is trained by learning all the bird names and their physical features; and second, it is a social activity because you do it as part of a group. The physical exercise in the fresh air is a further benefit. Again, more research is needed to find out whether these activities lead to permanent improvements in mental abilities or a long-term delay in decline.

Finally, there's this. A 102-year-old man from Belgium was invited in February 2012 to start at the local preschool the following term. It was, of course, an administrative error. But maybe, in the future, invitations like this will be sent, not for preschool but for a school for 70- to 80-year-olds. In fact, such schools already exist. The Seniors Academy in Groningen celebrated its 25th anniversary in 2011. Like similar institutions elsewhere, the Academy offers courses for seniors — with a minimum age of 50 — who want to develop their talents further. On the basis of the research we have looked at here, we can be sure that these types of courses are an excellent way to keep the mind sharp.

PHYSICAL EXERCISE

Last but not least, of all the ways to improve and protect our mental abilities, physical exercise has the strongest supporting evidence. As we saw earlier, the saying 'a healthy mind in a healthy body' has retained its validity over the centuries. But does this mean that promoting physical health through sport helps to stop the decline in our cognitive processes? And if so, how does it work?

Many older people get too little exercise. Individuals aged 55 and above are less active than the average person, and with age the difference continues to grow. According to the Department of Health, only one in four Australians over 75 is sufficiently active (as compared with 43 per cent of the adult population). The recommendation for people over 55 is half an hour of moderately intensive exercise on at least five days and preferably every day of the week. 'Moderately intensive' means that your breathing and heart rate increase, your body temperature rises, and you work up a sweat. It is indisputable that exercise improves health, but does it also benefit the brain?

Maaike Angevaren, a physiotherapist and researcher at the HU University of Applied Sciences Utrecht, contacted me in 2004. She had an ambitious goal: to write a general review of the effects of sport on older people's mental abilities. She wanted to submit the article to the Cochrane Library, the database of the prestigious Cochrane Collaboration. The Collaboration is an independent, not-for-profit international network of medical researchers, and it publishes pioneering reviews of all kinds of treatments. These reviews collect and integrate all available data from high-quality studies to establish whether or not there is conclusive evidence about a specific treatment. I thought it an excellent idea to research the literature on the subject and to work with her team, whose members included a sports physiologist and a geriatrician.

Maaike was primarily interested in aerobic exercise, in which respiration and heart rate increase. Moderately intense activity improves heart and lung function, while a good supply of blood and oxygen is vital to the functioning of the brain. What's more, a healthier heart means a smaller risk of a heart attack or brain haemorrhage. Maaike reviewed the literature and found 11 articles reporting on high-quality research into the effects of moderately intense physical exercise on cognitive function. In a typical example, subjects engaged in moderate-intensity workouts for half an hour three times a week for 12 weeks. They were compared with subjects in a control group who only did stretching exercises. Most of the studies found that the group doing the workouts showed an improvement in both heart/lung function and cognitive function, though not in all the neuropsychological tests administered in the study — it was primarily speed of processing and concentration that improved. Maaike herself observed that 'the brain can be compared to a computer and the time it needs to boot up. The older the computer, the longer it takes to get going.' [4] Training can speed up information processing in the brain. Incidentally, the fact that both heart/lung function and cognitive function improved doesn't mean that one leads to the other. Other factors that are influenced by this type of exercise may be involved, such as the release of growth factor into the blood, which has a beneficial effect on the brain. We'll come back to this later.

Does all that huffing and puffing in the gym or the park have an effect on our brain tissue? Yes indeed — the brain gets bigger. And that can't hurt, given the average amount (15 per cent) by which the brain shrinks between the ages of 30 and 90. In one study, MRI scans were performed to measure the volume of grey matter in the brains of 60 participants between the ages of 60 and 79 after they had taken part either in moderate-intensity or stretching exercises for six months. In the group who had engaged in moderate-intensity exercise — but not the other — the researchers observed an increase in grey and white matter in the front part of the brain. In another study, carried out at the University of Pittsburgh, Pennsylvania, 120 participants (with an average age of 67) were assigned to the same two groups, doing either moderate-intensity or stretching exercises. This time, their MRI scans were compared after a year, focusing on the hippocampus, which is essential to storing information in the memory. In the moderate-intensity group, the hippocampus had grown by 2 per cent, while that of the stretching group had shrunk by 1.5 per cent, the normal amount of shrinkage in people of that age. The researchers also looked at two other areas of the brain that were less affected by ageing, and found no differences. Thus, it is not true that every part of the brain is enhanced by moderately intense exercise, but the parts that decline as we grow older are. What makes these findings even more convincing is that memory performance improved as well: the greater the increase in grey matter in the hippocampus, the more memory improved.

The reason why exercise is so good for our cognitive function is clear: our grey cells get reinforcements. In fact, studies with large numbers of participants who exercise regularly show that this reduces the risk of Alzheimer's, sometimes by as much as 50 per cent. The best situation is where the individual has always been physically active, but even people who start when they are 60 can benefit. And an extra advantage is that doing sufficient exercise lowers the risk of chronic disease: it has positive effects on those diagnosed with high blood pressure, high cholesterol and blood-sugar levels, and obesity.

So how does physical exercise get those grey cells working better? First, by increasing the flow of oxygen to the brain; and second, by activating growth factors. Moderate-intensity exercise stimulates the release of growth factors into the bloodstream and brain. One of the factors is IGF, which, as we have already seen, promotes the healing of damaged tissue in the brain, the growth of neurons, and the formation of new connections between them. Another substance that scientists are very interested in is BDNF, or brain-derived neurotrophic factor, secreted by the brain to promote the growth of neurons. BDNF doesn't work only in the brain but elsewhere in the body too, where it promotes the recovery and growth of nerve tissue. The Pittsburgh study that I referred to earlier, which demonstrated growth in the hippocampus after a year of moderate-intensity exercise, also measured the concentrations of BDNF in the participants' blood. Higher concentrations of this factor were linked to an increased number of grey cells in the hippocampus. This finding correlates with results from animal research. In brief, the simplest way to keep your brain as healthy as possible is to walk, swim, cycle, or work out at a brisk pace for 30 minutes at least three times a week. That way, you don't need to go to your doctor or pharmacist, but you still get the best medication through the miraculous way your body works.

There are forms of exercise that require less effort than sport but can still be valuable. This is especially relevant for older people who have difficulty walking. Tai chi, for instance, is a traditional Chinese martial art. In its 'soft' form, participants make slow, flowing movements that involve contracting and relaxing the muscles and consciously controlling breathing. Though there is no scientific evidence to back the ancient Chinese beliefs regarding meridians (energy pathways), there are reliable research results demonstrating the health benefits of the movements themselves. People can practise tai chi even after reaching an advanced age. Still, more research into its effects on the brain is needed.

The same applies to the claim of the Amsterdam psychologist Erik Scherder (who is also a professor of human-movement sciences in Groningen) that chewing is good for the brain. There are indications from research that chewing gum improves the flow of oxygen to the brain and can lead to improved cognitive skills. Since many older people have difficulty in chewing, they tend to be given food that doesn't require much mastication. As a result, their ability to chew declines even further. Scherder thinks that good dental care for older people is essential, and that they should eat food that exercises the teeth and gums. And he thinks nursing homes should buy in large quantities of chewing gum!

- Physical exercise helps to keep the mind sharp, as do memory and concentration exercises. It seems obvious, but it isn't: many people over the age of 65 engage in too little physical activity and mentally stimulating pursuits.
- There is currently no medication that can reliably and safely improve our memory or other cognitive abilities, though new medicines are constantly being tested.
- Placebos have produced remarkably positive results in enhancing memory.
- Many preparations claim that they are good for the brain, but only in the case of vitamin B12 and omega-3 fatty acids is there scientific evidence that they can lead to improvements or delay decline.
- People who have been doing 'brain work' all their lives have a lower risk of Alzheimer's.
- People who exercise regularly have an even smaller risk of Alzheimer's. It's best to start young, but even people who only begin to exercise at the age of 60 can benefit.
- Older people who have difficulty walking but want to exercise should try tai chi.

7

With Age Comes Wisdom:

why older people are wiser

If someone asks me whom I regard as a wise person, I always think of Kofi Annan, former UN chief and winner of the Nobel Peace Prize. Though I have never met him in person, in the media he talks sense in a tactful and friendly way. In 2007, chaos followed disputed elections in Kenya. Official counts showed that the sitting president, Mwai Kibaki, had won 47 per cent of the vote while his rival, opposition leader Raila Odinga, had only 44 per cent. But these results were widely questioned. Followers of both leaders, who represent different tribes, were incited to attack each other by politicians on both sides. Hundreds of people died. Long regarded as one of the most stable and prosperous countries in Africa, Kenya attracted thousands of big-game tourists every year. Now it had become the scene of lynch parties and people fleeing in panic from their compatriots. The chairperson of the African Union, Ghanaian president John Kufuor, was quickly called in to try to bring the two sides closer together and work out a solution. He didn't even manage to get Kibaki and Odinga to the negotiating table. Then Kofi Annan, who was nearly 70, agreed to take over. He engineered a reconciliation between the two sides, and a coalition government with Kibaki as president and Odinga as prime minister was formed. Wisdom is certainly needed to bring people together in such extreme circumstances.

But even the wisest people cannot achieve the impossible. Even Kofi Annan was unable to convince the Iraqi dictator Saddam Hussein that he should co-operate with the UN Security Council. In February 1998, he spoke for three hours with Hussein, doing his utmost to win him over. 'In the end, the only means I have is reason and persuasion,' he said later. When negotiating with government leaders, Annan would try to see the situation through their eyes and find issues on which they could agree. This is an important element of wisdom: the ability, by using knowledge of the past and taking different viewpoints into account, to put forward proposals that may lead to something positive in the longer term.

In all cultures, wisdom is associated with experience and handing on knowledge gained in the past. During time spent on a Native American reservation in southern Arizona, the American psychologist Louis Cozolino met a tribal elder who called himself Mister John. 'His bright eyes were sunken within a patchwork of dark wrinkles,' wrote Cozolino. 'I estimated him to be about 80 years old.' In answer to a question about wisdom, Mister John told him that it consisted of the stories of his forefathers, passed on by 'the old ones'. These stories were a lifeline to their history. He believed that people needed a connection in order to live together, and on their reservation, that connection was the tribe. Young people need guidance to make the right long-term choices for themselves and for others, choices that mean they will be surrounded by love and friendship. The people around you are your life's riches.

Mister John's words are diametrically opposed to the slogan that appeared on posters displayed in physics departments in some American universities in the 1980s. It read: 'The knowledge of the previous generation is no longer valid.' Indeed, many scientific insights change over the years and new knowledge is acquired. But are wisdom and knowledge always the same thing? Young people today have often benefited from higher education and are at home with modern media. Before you know it, they've found the answer to a question on the internet. But if wisdom is about insight into complex life issues or dealing with difficult situations, the perspective that age brings might well come in useful.

WHAT IS WISDOM?

Throughout the ages, and in every culture, there have been people regarded by their contemporaries as repositories of wisdom. They were often grey-haired old men valued for their religious and philosophical knowledge and experience, which gave them insight into the major issues in life.

But how can a person be wise if their grey cells are shrinking and their memory and concentration going downhill? To answer that question, we should first define wisdom and see whether it does indeed increase as we age. If it does, we then have to reconcile that with the changes we have noted in the brain.

Scientific research always requires a definition. But because it is not easy to pinpoint what wisdom is, researchers tend to use different definitions. The following is perhaps a good place to start: wisdom is the ability to understand complex situations and thereby promote optimal behaviour, so that the outcome satisfies as many people as possible and takes everyone's wellbeing into account.[1] But this is not an entirely satisfactory definition. To try to establish what people generally believe wisdom to be, an Austrian and an American researcher developed a questionnaire that was completed by over 2,000 readers of *GEO* magazine. Many answers referred to the ability to understand complex issues and relationships, knowledge and life experience, self-reflection and self-criticism, acceptance of others' perspectives and values, empathy and love for humanity, and an orientation towards goodness. These perceived elements of wisdom are not unique to *GEO*'s readership; they are subscribed to worldwide. The American psychiatrists Thomas Meeks and Dilip Jeste added two qualities to the list: emotional stability, and the ability to make decisions in ambiguous situations. And, finally, there is humour. Though it isn't generally regarded as an essential component of wisdom, a sense of humour must surely be part of the self-knowledge required for true wisdom. Jeanne Louise Calment, a French woman who lived to the age of 122, was known for her wit. On her 120th birthday, a journalist expressed, somewhat hesitantly, the hope that he would be able to congratulate her the following year too. 'Why of course,' she replied, 'you look very young.'

Though people have recognised the importance of wisdom for thousands of years, it was, until recently, almost entirely absent from medical research into ageing.

This may be because Western culture puts greater emphasis on the intellect. Cognitive skills, such as memory, concentration, and logical thinking, have of course been extensively studied. But knowledge, skills, and expertise are not the same as wisdom, which has to do with a broader insight into life and making choices in ambiguous situations. It is also about achieving a balance between polar opposites: between weakness and strength, doubt and certainty, dependence and autonomy, transience and infinity. We consider people wise if they can give good advice in difficult circumstances and their judgements are balanced.

But research into wisdom doesn't have to be confined to living individuals. We can also look at what has been written on this subject over the centuries and in different cultures. In many cases, these are religious texts. The most well known example in the West is the Bible. In the book of Proverbs, wisdom is presented as more valuable than precious metals or jewels: 'Doth not wisdom cry? and understanding put forth her voice? ... Receive my instruction, and not silver; and knowledge rather than choice gold. For wisdom is better than rubies; and all the things that may be desired are not to be compared to it.'[2] The Bible also links wisdom to age: 'With the ancient is wisdom; and in length of days understanding.'[3] Christians believe that true wisdom is acquired in a personal relationship with God, who is the source of all wisdom. St Augustine distinguished between two kinds of knowledge: *sapientia*, or knowledge concerned with eternal reality (wisdom) and *scientia*, or knowledge of the natural or material world (what we call science).

Long before Augustine, the Greek and Roman philosophers who shaped much of our Western culture attached great importance to wisdom. Sophocles (5th century BCE), for example, wrote in his play *Antigone*: 'Wisdom is the supreme part of happiness.'

Eastern culture, too, has for centuries placed enormous value on wisdom. Its understanding of the concept has much in common with Western ideas. The *Bhagavad Gita*, written in India around the 5th century BCE, is a major work on this subject. It sees wisdom as the sum of life experiences, the ability to deal with emotions, self-control, love of God, compassion, humility, and self-sacrifice, all of which are also part of the Western concept of wisdom. The American psychologist Douglas Powell, who interviewed over 300 older people in the course of his research, calls humility 'a gift of experience'. During the course of their lives, older people have had moments of disappointment, failure, missed opportunities, and plain bad luck. These setbacks have been described by researchers and psychologists as 'wisdom-prone circumstances'.

Another influential work is the *Tao Te Ching*, a Chinese manuscript dating from around the 6th century BCE and one of the most important Taoist texts. It regards intuition and compassion — more so than reason — as the way to wisdom. Later on, the influential philosopher Confucius (551–479 BCE) called on his followers to improve the world, but warned them they had to start with themselves. 'To know what you know and to know what you don't know is true wisdom.' This concept of the limited nature of one's knowledge recurs in our current take on wisdom. And it is something you often see in older people as a result of their ability to see things in perspective. For instance, 73-year-old Joanne does not believe that her mental abilities have declined in the past few years. She points out that intelligence is not the same as wisdom, and that she has grown wiser in the course of her life. She now has more insight into the choices that she makes because she has a better understanding of the advantages and disadvantages that each entails. 'Nowadays I'm less certain that I always make the right decision,' she says, 'but that's not such a bad thing.'[4]

HOW OLDER PEOPLE REASON

The Swiss psychologist Jean Piaget (1896–1980) made an important contribution to our understanding of cognitive development in children. He described four stages, the last of which is 'formal thinking'. This begins around the age of 11 and continues into adulthood. A person who has reached this stage is capable of logical reasoning and can solve abstract problems; in other words, they can imagine various logical solutions to a problem and can test them through trial and error. The wrong solutions can then be eliminated, and what is left is the right one. Imagine, for example, that you're 12 years old and your mobile phone doesn't work. Your first reaction is that maybe the battery is dead; your second, that perhaps it's because you dropped it in some water yesterday. You test your first idea by recharging the battery, and if that hasn't worked after several hours, you conclude that the problem is likely to be water damage. Now, older people are no better than young adults at this kind of logical thinking, except perhaps in highly complex situations where past experience is helpful, such as applying for a mortgage. Yet in the case of new, artificial problems such as those created for psychological experiments, older people are actually at a disadvantage because they make heavy demands on concentration and working memory, precisely the things that deteriorate with age.

There is, nevertheless, a form of thinking that improves with age. Based on Piaget's terminology, behavioural scientists have come up with the concept of 'post-formal thought', which is used in confronting complex everyday problems that can be solved in different ways. It involves more uncertainty and more flexibility. To arrive at a solution, you often have to put yourself in others' shoes. In one experiment, subjects in varying age groups were presented with the problem of a student who had plagiarised large chunks of Wikipedia for a paper.[5] The student admitted she had taken whole passages from Wikipedia, but argued that she had never been told she had to list her sources or how to do so. The subjects were asked what steps they would take as a member of the examinations committee reviewing her case. The guide given to students clearly states that plagiarism is a serious offence that may be penalised by suspension or expulsion from the university. What was the result? Many young people thought that the student should be expelled. This is the outcome of formal thinking as described by Piaget. Their conclusion was logical: a rule had been broken, and the corresponding penalty should be imposed. Many of the older subjects, however, were less certain about their answer and applied post-formal thinking. Before they made a decision, they wanted more information. Was the student really unaware of the procedures? How far had she got with her degree? Had the concept of plagiarism been properly explained? Depending on the answers to these questions, older people might well have come to the same conclusion as their younger counterparts, but they were more likely to see things from the student's perspective and to consider the consequences of the penalty.

Another example of a problem where post-formal thinking works best is the dilemma confronting Harold and Hanneke.[6] Both recently turned 68, and will soon be celebrating their 40th wedding anniversary. They realise that they must think about the future and whether they should continue to live in the spacious home where their children grew up. Harold wants to sell the house and move into a luxury apartment in a complex for seniors that has lots of facilities, including a gym. The complex is also well served by public transport, which will be essential once they cannot drive anymore. Hanneke is not enthusiastic about the idea. She wants to stay in their home and undertake modifications when they become less physically able. For example, safety bars could be installed in the bathroom, and a large room downstairs could be converted into a bedroom. Harold and Hanneke think they need to talk through all the possible advantages and disadvantages of both courses of action with each other and their adult children. The fact that they intend to take their time over this decision, to consider all the pros and cons, and to weigh up each option shows they realise that a quick, logical answer is not available. They are thinking post-formally.

THE OLDER, THE WISER?

Does growing older automatically mean we will grow wiser? Not for all of us, unfortunately. In all age groups, there are people whose thoughts and actions cannot be described as wise, though this doesn't mean that they won't acquire more of the characteristics of wisdom as they age. Generally speaking, the experience of life and its ups and downs builds wisdom. But this is very difficult to measure. If you present people with complex situations and ask them for the best solution, many older people do no better than middle-aged people, according to a German study. What was interesting was that older people, like young adults, were better at solving problems that were typical of their age group. The subjects had to respond to a number of difficult situations — some of which involved a younger person, some an older. An example of the first was the story of Michael, a 28-year-old mechanic with two young children who hears that the factory where he works will close down within three months. There is currently no suitable work for him in the area where he lives. His wife is a nurse and has just started work again locally in a well-paid job. Michael doesn't know whether they should move to another town where he can find a job, or if they should stay where they are, with him taking on the role of stay-at-home dad. What would be the best solution for the next three to five years? What additional information is needed to come up with a solution? An example in the older category was Sarah, a widow of 60. Having recently completed management training, she has set up her own business, a challenge she has long wanted to take on. However, her son has just lost his wife and has two young children to care for. She could either give up the business and move in with her son to help out, or she could help him financially with costs of paid care for the children. What is the best plan for her for the next three to five years? What additional information is needed to solve her problem? Older subjects (aged 60–81) were better at devising a solution for Sarah, and the younger group (25–35) came up with better answers for Michael. In order to qualify as 'wise', the participants had to name various aspects of the problem, come up with a number of solutions, list the pros and cons of each, recognise uncertainty, assess risks, and finally devise plans for critical follow-up and possibly reconsideration of the chosen solution.

A small proportion of seniors will not perform as well as middle-aged people on this type of task, where solutions to complex problems are required. This is because it draws heavily on cognitive abilities such as working memory and executive functions (the ability to plan and empathise, for example). Older people whose skills have eroded more than the average will find it more difficult to think up a range of solutions and to compare them with one another. Though intact cognitive functions don't necessarily lead to wisdom, they do make a contribution. You can still be wise if your mental faculties have declined, especially in situations that are familiar to you from experience. But when faced with new problems that require weighing up a lot of information, a decline in working memory and cognitive flexibility tends to work against you.

THE TORTOISE AND THE HARE

In 2004, neuropsychologists at the University of California described a patient they called 'a modern-day Phineas Gage'. The reference was to a 19th-century railway worker of that name, one of the most well-known patients in the history of neuropsychology. The brain damage he suffered taught us much about the function of the hitherto mysterious frontal region of the brain (the prefrontal cortex). In 1848, Gage was involved in a dramatic accident: following an explosion, an iron bar penetrated his face with tremendous force and exited through the top of his head. To the amazement of his colleagues, he survived the accident and was even discharged from hospital two months later. But he was a changed man: as a close friend put it, 'Gage was no longer Gage.' Though his powers of reasoning, observation, and memory were intact, his personality was radically changed. A man known to be hardworking, energetic, and possessed of managerial abilities had, it is said, become impatient, foul-mouthed, and lacking in empathy. He was no longer able to assess the emotional significance of situations and could not control his own emotional reactions. He was subject to regular outbursts of rage, and was unable to plan his activities. Reconstructions of his brain based on his preserved skull show that it was primarily the underside of the prefrontal cortex that suffered severe damage.

The modern-day Phineas Gage in the case reported in 2004 had suffered damage to his head in 1962 when his Jeep hit a landmine on a military mission abroad. As a result of the explosion, the metal frame of the windshield penetrated his forehead. Like Gage, he appeared at first to have suffered no damage to his mental faculties when he left the hospital. He performed well on neuropsychological tests, and his intelligence was intact. Socially, however, he was not functioning well. He displayed disinhibited behaviour and an inability to control himself, both of which led to problems in his interactions with others. He lost his job, broke up with his wife, and became estranged from his children. According to geriatric psychiatrist Dilip Jeste, damage to the prefrontal cortex seems to lead to the opposite of wisdom: impulsiveness, socially inappropriate behaviour, and emotional awkwardness. Together with colleagues, Jeste mapped out for the first time a network of brain regions that are essential to wisdom. They attributed a special role to the prefrontal cortex.

Neurologist Elkhonon Goldberg proposes the same thing in his book *The Wisdom Paradox*. He sees the prefrontal cortex as the conductor, and the other regions of the brain as the orchestra. The prefrontal cortex doesn't make the music, but co-ordinates, integrates, and directs. This is why people who have suffered damage to the prefrontal cortex are still perfectly capable of performing many tasks, but encounter problems (or have delayed reactions) when things become complex, as is the case in social situations. Goldberg pointed to two other functions of the prefrontal cortex. First, our ability to empathise; and second, to activate certain sequences of actions, especially those involved in complex situations. For example, if you have years of leadership experience, you automatically know what steps to take in many situations. Goldberg gives the example of Winston Churchill, who suffered occasional mental lapses, but remained a brilliant leader despite his age.

There are four areas of the brain essential to wisdom (see figure 20). First, the ventromedial prefrontal cortex, which plays a role in emotional responses and decision-making. Second, the outside of the prefrontal cortex (technically speaking, the dorsolateral prefrontal cortex), which is involved in rational thought and determining a strategy to solve problems. Third, the anterior part of the cingulate cortex, which detects conflicts between competing interests and between rational processes and emotions. And finally, lying deep in the brain, the striatum, which is the brain structure activated by stimuli associated with reward.

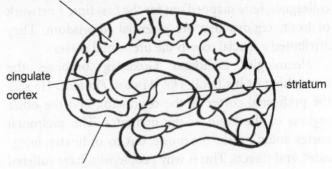

FIGURE 20. Areas of the brain that are essential to wisdom: the ventromedial prefrontal cortex (in the frontal lobe at the bottom), the dorsolateral (or the outside of the) prefrontal cortex, the cingulate cortex, and the anterior (front) part of the striatum.

Research has provided indications that older people are more focused on the rewards that follow good solutions than on the negative consequences of mistakes, which means that they place an emphasis on achieving constructive answers rather than preventing errors. If you want to teach a 75-year-old to use a computer, it's better to concentrate on what they do well, rather than constantly pointing out their mistakes or reminding them to do things differently. While it's fine to occasionally say, 'Hang on, that's not the right way!' when teaching a younger person a new task, it's the wrong strategy where an older person is concerned. This could be the result of changes in the way that certain areas of the brain function as we age: the anterior cingulate cortex, involved in error detection, is less rapidly activated (in many people, the number of grey cells in this area declines as they age) while the structures forming the 'reward system' remain unaffected.

Using an electroencephalogram (EEG) to measure electrical activity along the scalp, a team of German researchers found that a spike in brain activity occurred in young and middle-aged people when they were informed that they had made a mistake. This spike represents activity in the cingulate cortex. The more elevated the spike was (and therefore the stronger the activity in the brain), the faster the person learned from the mistake. But in older subjects, the spike was much weaker. Older people use other regions of the brain in learning, primarily the prefrontal cortex, which is essential to working memory. Though the function of this area of the brain also declines slightly, many older people manage to get the best out of it. They do this partly by mobilising extra brain activity, as we saw in chapter 2.

In general, older people have more difficulty with completely new tasks than with ones in which they can fall back on existing knowledge gained through experience. The well-stocked 'database' that they have built up over the years helps them resolve many routine issues with ease.

Dr Oury Monchi of the University of Montreal in Canada likes to refer to one of Aesop's fables when explaining the results of his research into the brains of elderly people. In the race between the tortoise and the hare, it is the tortoise that wins, even though he is much slower. He knows how to make the best use of his abilities, while the arrogant hare takes a nap during the race. Monchi and his colleagues asked a group of older and a group of younger adults to classify words into categories while in an MRI scanner. Words could be categorised according to rhyme, meaning, and their initial letter, but the researcher constantly changed the rules without informing the subjects. While classific-ation according to rhyme ('door' would belong with 'floor') was initially correct, it suddenly became wrong and the subjects had to decide whether they should now categorise by meaning ('door' would belong with 'house'). The older subjects, unlike their younger counterparts, did not show increased brain activity in response to negative feedback ('Wrong!'). Rather, they exhibited increased brain activity primarily when they had to make new choices about the words. In other words, they appeared to invest more in thinking up new strategies to perform the task, which is a more active response than simply reacting to error warnings.

THE EXPERIENCED DECISION-MAKER

Many psychological experiments are artificial: the tasks they set have nothing to do with daily life. In this type of artificial test, people who are around 20 perform better than those around 70. An example of this is where the subject repeatedly has to choose one of four cards shown on a computer screen. Because a different (bigger or smaller) reward is attached to each card, subjects can discover which strategy offers the most benefit. Younger people are often better at this than their older counterparts.

In this test, each decision is isolated: earlier decisions have no effect on later decisions. But in daily life, one decision often has an impact on another, and choosing short-term benefits may not always lead to long-term gains. Imagine you'd really like to take a holiday in France, but it's too expensive. So every year you go to New Zealand. But you could do without a holiday one year and save the money to go to France the next. The first part of the decision (no holiday one year) is less rewarding, but the next part (achieving your dream of visiting France) more than compensates. Or you decide to buy a more expensive washing machine, which produces no immediate benefit in comparison with a cheaper model, but may in the long term save money in terms of energy consumption, repairs, and the machine's longer lifespan.

With this idea in mind, psychologists at the University of Texas performed an experiment based on realistic scenarios in which one decision affected subsequent decisions. The best possible strategy could only be adopted if the subject had a long-term vision. Here, older subjects performed better than younger people. It is not that young people cannot do this, but it is remarkable that older people can do it better, given that their executive functions (working memory, ability to compare, cognitive flexibility) deteriorate as they age. Again, experience may be an important factor.

Research has also demonstrated that older people take fewer risks in financial decision-making and are less impulsive. This is to do with their greater use of both hemispheres. In ambiguous situations where the risk of losing is greater than that of winning, this decision-making style is more sensible than a risk-taking approach. Nevertheless, it is sometimes necessary to take risks in order to progress. To get the best of both worlds, it might be wise for financial organisations and investment banks to employ people over 65 alongside their younger staff.

As we have seen, not all our mental faculties go down-hill as we age. Many of them (including language skills, logical reasoning, general knowledge, and spatial insight) hardly decline at all. And older people are even better than their younger counterparts in solving certain kinds of problems. They can call on a larger database of knowledge and experience, and can make complex decisions relying on intuition.

Paradoxically, this may — at least partly — be the result of the fact that older people's brains work more slowly, so they react less impulsively. And because it takes longer for them to arrive at a decision, they have more information on which to base it. According to Louis Cozolino, you might expect three phenomena in wise people, neuropsychologically speaking: more diverse activation of different areas of the brain, slower processing of information, and integration of cognitive and emotional functions. And that is precisely what the ageing brain is like. There is more diversity in the use of brain structures, because older people can draw on a greater store of knowledge amassed over the years and have learned more ways to tackle a problem. We have seen that they are more likely to use both sides of the brain, which then work together to achieve good results. Cozolino's second point, on slower inform-ation processing, is inevitable given the fact that the brain's highways (the white-matter pathways) break down as it ages. The advantage of this, on the other hand, is that fewer hasty conclusions are reached. And better integration of cognitive and emotional func-tioning is the result of older people learning to assess

the value of both reason and emotion, and to give each
its own place.

- Wisdom can be defined as having insight into
  the major issues of life and the ability to make
  balanced decisions in uncertain situations.
- We become wiser thanks to the erosion of our
  mental faculties. The ageing brain works more
  slowly, and our responses are therefore more
  sensible.
- Older people can draw on a substantial database
  of knowledge and experience, and can use their
  insight to arrive at complex decisions.
- Older people take fewer risks in financial decision-
  making and are less impulsive. This is the result of
  making more use of both sides of the brain than
  when they were younger.

8

The Best Possible Brain:

scientifically proven advice

On 26 November 2011, his 100th birthday, Robert Marchand established a new world record in the French commune Mitry-Mory. He cycled for an hour in the gym, covering 23 kilometres. He had decided not to cycle on the road as it was raining and he was afraid of falling. Three months later, he repeated his performance, this time on a real racing bike and covering over 24 kilometres, in front of a group of supporters at the International Cycling Union velodrome in Aigle, Switzerland. This achievement was registered by the World Record Academy. Considering that an average cycling speed is 12 kilometres per hour, Marchand's score is remarkable. In an interview afterwards, he joked that he had deliberately taken it easy so that others could break his record. Another reason that he didn't push himself was he had promised his doctor not to allow his heart rate to exceed 110 beats per minute. Not that there was a problem — the week before, he had undergone his first electrocardiograph, and it had shown no serious irregularities. When asked what magic potion he had in his water bottle, he replied triumphantly that his 'doping regimen' consisted of water with a dash of honey. His recipe for success? He pointed out that he had always led a healthy life, had never smoked, and drank little alcohol. When he was just shy of 90, he took part in the Bordeaux–Paris race, cycling 600 kilometres in 36 hours.

For a centenarian, Robert Marchand is remarkably fit. He still lives alone and goes for a bike ride almost every day. In 2011, a mountain pass in the French department of Ardèche was named after him. His advice to young and old? Keep moving!

SUCCESSFUL AGEING

Because there are more older people today than ever before, the question of how to keep your mental faculties as sharp as possible is highly topical. Sigmund Freud believed that our cognitive abilities lose their elasticity after the age of 50. In other words, you can't teach older people anything new. But now we know the brains of senior citizens have a degree of plasticity, which means that neurons can make the new connections needed to learn new things and can compensate for the erosion of brain structure and function as we age. Of course, whether someone grows old 'successfully' depends on how you define success. In the scientific literature, it is determined by three things:

- an absence of chronic illness and impairments
- an ability to function mentally and physically at a reasonable level
- a strong connection with social contacts, family, and friends.

How many 65-year-olds fall into this category? It depends on the criteria you use to define each factor. Some researchers estimate it at 5 per cent, which is unsurprising if you take the phrase 'absence of chronic illness and impairments' literally — that excludes almost everyone. Consequently, most studies only count illnesses that last for a long time, recur frequently, and impede function. Taking this and the other two factors into account, a common estimate is 30 to 50 per cent of older people in Europe and the United States.

The German psychologist Paul Baltes argued that any assessment of whether people are growing old successfully should take the views of those people into account. And what is important is the extent to which they adjust to the limitations that ageing brings. A 60-year-old can't hit a tennis ball as hard as he could when he was 25, but that doesn't mean he can't enjoy the game anymore. He might now focus more on position and strategy to compensate for a loss of speed and strength. Thus, what matters is not one's declining abilities, but whether a person manages to cope with these changes in such a way that his or her quality of life isn't seriously affected. Compared with younger adults, older people have more physical ailments, which can lead to depression. Yet they are no more likely to be depressed. How you deal with setbacks is the important thing.

Growing old healthily has to do with hereditary factors, but that's not the whole story. Research has shown that genes account for only about one-third of the impairments that accompany ageing. More important (and accounting for two-thirds of impairments) are environmental factors — lifestyle, social support, medical care, and so on. In this chapter, I list all the recommendations for optimal care of the ageing brain. A decline in cognitive function is a normal part of growing older. But a healthy and active lifestyle, as we saw in chapter 7, can considerably slow that decline. Though Robert Marchand has probably been blessed with a strong constitution and a passion for sport all his life, his habits and daily activities have undoubtedly contributed to his extremely healthy old age.

CHOOSE YOUR PARENTS CAREFULLY

In *Ten Commandments for the Brain*, Utrecht professor of psychiatry René Kahn gives ten guidelines for a healthy brain, which apply equally to the ageing brain — for example, 'make friends', 'make music', and 'sweat' (engage in physical activity). But he also includes one recommendation that is difficult to follow: 'choose your parents with care'. Obviously, this is intended to underline the importance of hereditary factors. These explain, to a considerable degree, why some people grow old healthily, with few ailments and with their cognitive abilities intact, while others are plagued by diabetes, heart problems, and a failing memory. Even if you observe all the rules for healthy ageing, you may nevertheless experience serious decline — in which case there's a substantial chance that this is due to your genetic make-up, and nothing you do will be sufficient. The converse is also possible, of course. There are people who break all the rules and still retain their memory and powers of concentration. But most people will benefit from the advice that follows, which is all based on medical research.

Incidentally, when speaking of heredity, it's worth remembering that it doesn't work in isolation but interacts with all kinds of environmental factors. This interplay of factors was illustrated by a study involving 1,140 people between the ages of 50 and 70 carried out in Baltimore. The researchers looked at whether a particular gene — APOE, a strong predictor for Alzheimer's disease and mental decline — had an effect on the association between the neighbourhood where people lived and their cognitive function. With regard to the neighbourhood, the question was whether living in a disadvantaged area (with high unemployment, poverty, abandoned houses, crime, and drug problems) was linked to a deterioration in concentration, memory, and cognitive flexibility. This proved to be the case, but only in people who carried a specific variant of the APOE gene. Living in these neighbourhoods did not have a negative impact on cognitive function in people carrying another variant of the gene. It would appear that having the gene makes a person more susceptible to the stress of living in such an area, and what it comes down to is a complex interplay of influences. Many other studies have confirmed that in combination with environmental factors, such as social support, our genes can affect our brains in a way they would not normally do. How they interact is largely unknown, but this will undoubtedly become clearer as research progresses. Thus, in the future, people may be advised to move, or to ensure they have more social support, on the basis of genetic testing.

The study of external influences on the expression of genes (that is, whether they are turned 'on' or 'off') is known as epigenetics. There is currently enormous interest in this field, and major discoveries are being made. It has produced convincing evidence that our genes do not determine everything, and that we are not just our brains. We are who we are as the result of interacting with our environment — what we eat, where we live, whom we live with.

KNOW WHAT YOU EAT

The oldest and healthiest people in the world live in Okinawa Prefecture, a chain of subtropical islands in the East China Sea between Japan and Taiwan. For a long time, it was an independent kingdom, but it has been part of Japan since 1879. The Okinawans have an average life expectancy of 85, four years longer than in the UK. While Japan has the largest population of centenarians in the world, there are three times as many of them (per 100,000 inhabitants) in Okinawa than in the rest of the country. But Okinawans don't just live longer — they are also healthier, experiencing relatively few age-related diseases. Heart problems, strokes, and cancer are less common. This may, of course, be due to heredity. But genetic research has shown that this is only partly the case: Okinawans who migrate to another country live less long and are less healthy than those who remain on the islands. Thus, environmental factors are decisive.

Bradley and Craig Willcox are two American brothers who have conducted research into ageing in Okinawa for decades. (Bradley trained in internal medicine and geriatrics, and Craig is an expert in cross-cultural gerontology.) According to them, five factors play a role in longevity: the right diet, a lack of stress, a caring community, a high level of physical activity, and a spiritual commitment. Now, you might think that living a stress-free existence on a beautiful island with warm temperatures, fine beaches, and magnificent flora and fauna wouldn't be too difficult, but life isn't always easy for the Okinawans. Prices are high, as they are everywhere in Japan; and incomes are low, as there is little industry. Nevertheless, most of them are content — their calm and down-to-earth disposition certainly plays a role in successful ageing.

Okinawans are positive by nature and do not view growing old as a negative experience. We have already seen how important this kind of attitude is. They are friendly and peaceful, and it's hard to imagine that Okinawa is the birthplace of karate. But it's not the aggressive form seen in films: here, it is a form of self-defence based on control of body and mind. The slogan above the entrance to the karate school belonging to Fusei Kise, an 81-year-old grand master who still gives karate lessons at the US military base on Okinawa, reads 'Softness overcomes hardness'.

Okinawa's main secret is diet, say Bradley and Craig Willcox. It contains little saturated fat, sugar, and salt, and consists mainly of fish, shellfish, tofu, seaweed, rice, vegetables, and fruit. Okinawans drink a lot of black and green tea; both contain antioxidants and are good for health. An example of a recipe adapted to Western tastes (and availability of ingredients) features scallops, Chinese cabbage, breadcrumbs, soy sauce, olive oil, a pinch of sea salt, and low-fat mozzarella. After the scallops and cabbage have been fried, the whole thing goes in the oven for 10 minutes.

A crucial element of a brain-friendly diet is moderation: avoid over-eating. That, too, we can learn from the Okinawans. Calorie reduction reduces the damaging effects of oxidative stress on the neurons in the brain. Calories are energy, and energy drives what is called oxidative metabolism (the part of the metabolic process in which cells break down molecules into energy). The less fuel there is for the 'energy engine', the less damage to the mechanism. And reducing calorie intake doesn't mean you have to starve yourself. Just limit snacks such as biscuits and sweets between meals, and eat reasonable portions at mealtimes.

The diet of the average Western person contains too many carbohydrates and sugars. In combination with little exercise, this is a risk factor for diabetes, which in turn is bad for the brain and can hasten dementia. Of course, not every older person with diabetes will suffer from dementia at some point, but there is a higher risk of this.

It is also wise to reduce the intake of saturated fats, as they raise cholesterol levels. Too much cholesterol can lead to clogged arteries and increase the risk of a heart attack. Omega-3 fats, on the other hand, are beneficial.

A balanced diet includes proteins, grains, vegetables, and fruit. According to a study conducted at the University Medical Centre Utrecht, a class of phytoestrogens known as lignans are linked to improved cognitive function. Sources of lignans include sesame seeds, linseed oil, broccoli, cabbage, peaches, and strawberries. The most important phytoestrogens are lignans and isoflavones. Soy is rich in isoflavones, and as we saw in chapter 5, it has little impact on cognitive functioning. Lignans, on the other hand, may have a stronger impact.

Make sure you get enough vitamins, especially B12 and folic acid. The Department of Health in Australia recommends a daily intake of 2.4 micrograms of vitamin B12 and 400 micrograms of folic acid.

Finally, something that's often forgotten: keep your fluid levels up. Many older people drink too little. The importance of fluids hit the headlines in 2003, when thousands of elderly people died in France during a heat wave. An Australian study that used brain scans to map brain activity and satiation during the process of drinking found that the signals that activate thirst decline as we age. In young people, the process works differently: they have to drink more before their brains are satisfied and tell them they are no longer thirsty. The researchers concluded that the thirst signal only reaches the older brain if it is extremely strong, so they advised seniors to drink fixed amounts at set times of the day. The current guidelines are 2.6 litres of fluid a day for men, and 2.1 litres of fluid a day for women, most of which should be plain water.

KEEP ACTIVE

Zelma is 88 years old and has lived in Missouri all her life. She participated in a study of ageing conducted at the University of Missouri, in which the researchers identified her as a 'successful ager'. She has two married daughters, five grandchildren, and five great-grandchildren. She was married for over 50 years and has been a widow for the last 12. Two years ago, Zelma underwent surgery for breast cancer, after it was discovered through a routine mammogram. She is a retired primary-school teacher and still helps out at a local school two days a week. She is a member of two book clubs and the president of the local women's association, and is active in politics. A Democrat, Zelma was county co-ordinator of governor Russ Carnahan's campaign when he was running for office. She is also a Sunday-school teacher at her local Baptist church. Her hobbies are chair caning and quilting. An avid reader, Zelma enjoys reading the Bible — she's currently on her eighth reading, from beginning to end — and, after a cataract operation on both eyes, she can now see without problems. Her church and family mean the most to her, followed by her neighbours and friends. Asked if there is anything 'bad' in her life, she responds: 'Nothing — except, of course, that there is never enough time to do all the things I want to do.'

Perhaps this list of what Zelma does is just a little overwhelming. Don't worry — doing slightly less is also fine! But her mantra ('keep busy') is significant. It is exactly what all successful agers say: don't sit in a corner thinking about your health problems; be enterprising.

The importance of remaining as active as possible applies to all three domains — physical, mental, and social. We have established that physical exercise is good for the ageing brain. The specific recommendation was 30 minutes of exercise that makes your heart beat faster and your body sweat, at least three times a week. Of course, exercise can mean walking, cycling, swimming, or ball games — in fact, anything that increases your heart rate. If you're an older person, you do have to be careful not to become too fanatical: like Robert Marchand, make sure your heart rate doesn't exceed 110 beats per minute. For some, physical impairments may prevent you from exercising. In that case, some gentle movement is better than inactivity, and this is where activities that can be done in the bedroom or living room, such as tai chi, come into their own.

Working past retirement age is a good way to remain mentally active. Increased life expectancy and more years of healthy living mean that people will be working until they are 67 or even 70. But if you have retired, it's good to have hobbies that make demands on your concentration and memory: the history of local politics, architecture, or the sports association, for example; or a course of study. You could organise a book group or a debating club. But working with your hands also requires concentration and memory. Building a doll's house for your grandchildren, for instance, means coming up with a design and perhaps researching the subject on the internet. One of the best ways to keep your brain active is by reading books. This is better than reading newspapers, where the articles are short and the gist of the piece is often in the headline. A book requires you to retain and integrate much more information in order to follow the plot. Even novels that are easy reads can help. Watching TV, on the other hand, has no beneficial effect.

Finally, it is important to remain socially active. This can become increasingly difficult for people over 75, as their peers begin to pass away. But most places have clubs and associations where they can meet people. The important thing is to take the initiative, not to wait for someone to drop in.

SPIRITUALITY, THE ART OF LIVING, AND MINDFULNESS

Researchers are taking an ever-closer look at the role of spirituality in successful ageing. There is no agreed definition of what spirituality is, though you could loosely describe it as being aware of a transcendent dimension to human experience. Spirituality can become more important as people grow older, as many biographies have shown. Examples of people who became interested in spirituality as they aged include the composers Johannes Brahms and Franz Liszt. Late in life, Brahms wrote one of his last great works, a cycle of four songs (*Vier ernste Gesänge*) imbued with the Lutheran spirituality that became increasingly important to him in the 1890s, when he was about 60. Liszt was a Roman Catholic and was even ordained as a secular clergyman in his fifties. His music from that period is evidence of how his spirituality deepened, though it was combined with an interest in the new.

Spirituality often has a religious dimension, but that needn't always be the case. Feeling one with the natural world or losing yourself in art or music is also a spiritual experience. But religious spirituality usually involves the development of a personal relationship with the sacred or transcendent.

Both spirituality and religion are beneficial to mental health in ageing. Studies in the US and Europe have shown that church attendance, reading the Bible, and prayer make older people less vulnerable to stress and depression. In other words, the subjects of the research were more comfortable in their own skins. Analysis demonstrated that this could not simply be explained by the social contact between members of a religious community; it had to be related in some way to the religious practices themselves. The researchers suggested that the feelings of inner peace promoted by religious ritual and the ability to place life experiences in a broader perspective might play a role.

So how do you construct a framework that gives meaning to your life? According to Lord Coggan, who became Archbishop of Canterbury when he was 65, it entails living your life in the fulfilment of a task given by God. Julia Burton-Jones quotes him as follows:

It is said of Jesus Christ that he knew that 'he
came from God and was going back to God'.
Between these two poles his life was lived. He
came from God; therefore God had a purpose
which he must fulfil. He was going back to
God; therefore life was moving towards a goal,
a climax … Life seen like that has a dignity
and a significance all its own.[1]

After his retirement at the age of 71, Coggan remained
active in the service of the Anglican Church for many
years. 'The joy of being a priest,' he once said, 'is that
your work never ends until they carry you out. Then
another begins — that's elsewhere.'[2] Donald Coggan
died in 2000 at the age of 90.

The American psychiatrist Dan Blazer, who conducted a study into spirituality in ageing, pointed out that
many religious people would object to the view that faith
is a means to promote your own health. Spirituality and
religion are not primarily meant to keep you healthy —
they have intrinsic value. Nevertheless, human wellbeing
is their central concern. Even setbacks, generally associated with reduced wellbeing, can through spirituality
lead to a more profound self-knowledge and help people
to put greater value on the everyday or give others more
support and understanding.

Related to spirituality is the concept of 'the art of
living'. According to Professor Jan Baars, who lectures
on ageing at the University of Humanist Studies in
Utrecht, the Roman philosopher Seneca offers wise
advice on this subject. Seneca came to the conclusion
that we experience life as being short because we allow
time to be stolen from us. For him, the art of living is in
essence a question of maintaining control over how you

spend your time. One way of doing this is to concentrate on the present. The main obstacle to happiness is constantly postponing it to some future date, so that you are unaware of the riches of today. This is not to say that the past and the future are unimportant, but they must be given a new significance based on the present. And the ability to come to terms with the past is, according to the philosopher, a store of treasure that can be added to as we grow older.

In recent years, psychology has focused increasingly on mindfulness, a practice derived from the Buddhist tradition whose aim is to foster emotional wellbeing. It uses certain Eastern meditation techniques, but also reflects Seneca's advice to concentrate on the here and now. In the practice of mindfulness, you consciously focus your attention on your current thoughts, feelings, and surroundings. Openness is the key: you mustn't make judgements, but simply observe from a distance. This attitude comes naturally to some people. They can usually cope better with their own emotions and those of others, and are less likely to suffer from depression. At the NeuroImaging Centre in Groningen, my colleagues and I studied brain activity in students who had just scored either high or low on a questionnaire asking whether they regularly adopted a conscious, attentive attitude in daily life. In the subjects who claimed to have such an attitude, their prefrontal cortex was more active and their amygdala (which helps process emotions) less active when they tried to reduce negative feelings. These feelings were stimulated through pictures depicting unpleasant situations, such as one showing an open leg wound. In people who had a 'mindful disposition', another area in the middle of the prefrontal cortex was more active. This area was involved in the conscious experience of emotions. The pattern of brain activity thus shows that the subject allows the emotion to be felt but retains some control over it. This may be the best way of coping with emotions — you don't allow them to overwhelm you (activating only your amygdala and not the area in the prefrontal cortex), nor do you simply repress the emotion (activating an area on the outside of the prefrontal cortex). In other words, you're open to your emotions but you're in control of them.

But what if mindfulness doesn't come naturally to you? The answer is training. Therapy for depression uses mindfulness training, but people without psychological problems can benefit from this, too. The idea is to train the ability to take an 'unprejudiced' look at situations, so that you experience everyday sensations in a more conscious manner and accept the events that occur. This makes people more stress-resistant. What is interesting is that it also improves cognitive flexibility, or the ability to switch to another way of thinking, which is frequently needed in daily life. Cognitive flexibility declines as we age, and mindfulness might correct this to some extent. Research has also shown that mindfulness encourages a more positive attitude to ageing.

Mindfulness helps us to cope with unpleasant emotions and experiences. But it's also important to focus on experiences that give us satisfaction, to be aware of the good things in our lives. Older people do this more than their younger counterparts. Experiments have shown that seniors look longer at a happy face than one expressing fear or sadness. A recent study found that when subjects looked at a happy face, an area in the prefrontal cortex involved in regulating emotion was more active in emotionally stable older people than in younger people and seniors who are less stable (and therefore at greater risk of depression).[3] The researchers believe that the ability to focus on positive information can be trained and can improve the wellbeing of elderly people. It is particularly important for this type of training to be developed for those seniors who are at greater risk of depression because they are emotionally unstable.

A FIVE-POINT PLAN

Our brain undergoes changes in ageing, and we all have to deal with them sooner or later. Some of these are impairments, comparable with the physical problems that develop as we grow older. We have seen that concentration, memory, and flexibility when encountering new information suffer the most. Slower thinking and observation are the main causes of this. Others include the reduction in grey matter as a result of shrinking neurons, and minor damage to white matter (which is vital to the speed with which stimuli travel from cell to cell). If someone's memory is clearly deteriorating in a way that is obvious to relatives, and this is confirmed by neuropsychological tests, they may be suffering from Alzheimer's. This involves serious brain damage, unlike normal ageing processes.

But ageing is not only a question of decline. The older brain is an experienced brain that seeks new solutions where necessary. If one side of the brain starts to have problems in processing a large amount of new information, the other takes over. We have seen that people actually get better as they age in making complex decisions and dealing with emotions, both important cognitive skills. Wisdom increases, too. In our Western culture, wisdom has been overlooked because of our obsession with youth. It's high time for a re-evaluation. An active lifestyle, a healthy diet, and an awareness of spirituality can help older people to keep doing the things they consider important.

Take the example of a neuroscientist and her own ageing brain. In 1960, Marian Diamond was the first woman to be a faculty member at the University of California, Berkeley. In 1974, she became professor of anatomy. She was at the front line of brain research for decades and examined sections of Albert Einstein's brain. Diamond developed a five-point plan to keep the brain young: diet, exercise, challenges, newness, and tender loving care. By challenges, she meant engaging in activities that demand effort, intelligence, or creativity. She advised looking for new stimuli — reading something different, or visiting new places and meeting new people. Diamond follows her own recommendations, and it hasn't done her any harm. At 81, she was still giving anatomy lectures at the university. When asked what her challenges were, she replied, 'Every student who sits in that chair [in her office in the Life Sciences Building]. They come in here asking questions, and you better have the answers.' And what about newness? 'I have grandchildren,' she said, adding that there was nothing better than inventing new things with them to stimulate their brains. But her feeling is that diet is the most important element of her plan.

Though I respect Marian Diamond's wisdom, my choice would be different. The strength of the ageing brain, as we have seen in this book, is understanding emotions and social situations on the basis of a lifetime of experience. Love and empathy are essential to good relationships with others, especially those closest to you. Even if you eat well and continue to feel fit at 80, without social contacts, family, or friends you can still be lonely. Diet, exercise, challenges, and newness are all important. But the most important of all is love.

IMPORTANT INSIGHTS

- Remember, we are not simply our brains. We are who we are as the result of interacting with our environment: what we eat, where we live, and whom we live with are all of decisive importance.
- The oldest and healthiest people in the world live in Okinawa Prefecture. They eat little saturated fat, salt, or sugar.
- The crucial element of a brain-friendly diet is moderation.
- Make sure you drink enough fluids, preferably 1.5 litres of water per day.
- The theory that physical exercise is good for your mental faculties has the strongest supporting evidence.
- Reading books is an excellent way to keep your brain active.
- Spirituality, religion, and mindfulness have a proven positive influence on mental health.

NOTES

CHAPTER 1

1 Aldo Ciccolini (born 15 August 1925) is an Italian–French pianist. He began his career at the age of 16. In October 2013, he was still giving concerts, and he was also scheduled to perform in Osaka on 26 and 27 June 2014. The review cited here was by Christo Lelie in *Trouw*, 12 May 2011.

2 The examples of Aristotle and the ichthyologist were cited in Salthouse (2012).

CHAPTER 2

1 Donald Sturrock, *Storyteller: the authorized biography of Roald Dahl*, Simon & Schuster, New York, 2010.

2 Gerben van Kleef, *Op het gevoel* (*The Power of Emotions*), Atlas Contact, Amsterdam, 2012. Van Kleef shows how emotions help us to understand others and further our own interests. We constantly influence others, consciously or otherwise, through our emotions. In turn, others influence us. The idea that emotions are irrational is outdated.

3 More information about cortisol and the effects of stress on the ageing brain can be found in *Het vitale brein* (*The Vital Brain*) by Eddy van der Zee (Bert Bakker, 2012).

4 The responses of the participants in the 2010 study by Grossman and others are paraphrased here.

CHAPTER 3

1 There are also indications that a slower metabolism has a positive effect on the brain. Italian researchers have discovered that a calorie-restricted diet promotes the activity of a protein called CREB1. This regulates a variety of brain functions, particularly learning and memory, and it becomes less active as we age.

2 Led by British professor Ian Deary, researchers examined the brains of 132 people aged 74 (Penke et al., 2010). Damage to white-matter tracts was linked to reduced speed of processing but not to problems with memory or executive functions. This kind of damage can be measured using a special MRI scanning method called diffusion

tensor imaging, or DTI. The technique probes the mobility of water protons in tissue. Because these protons prefer to move parallel to an axon rather than perpendicularly across it, researchers can use DTI to map the network of white-matter tracts.

3 Mainly chemokines (a type of signalling protein).

4 This pattern is known as HAROLD: Hemispheric Asymmetry Reduction in Older Adults.

CHAPTER 4

1 In this book, Strauch writes about developments in the brain and mental abilities in middle age (which she defines as 40 to 68 years old).

2 I conducted this literature search with Shankar Tumati, researcher at UMCG's NeuroImaging Centre. MRS measures the concentrations of certain substances involved in the brain's metabolism, known as metabolites. Small differences in the resonance frequency of hydrogen nuclei and groups of these nuclei in different molecules and at different places in molecules make it possible to identify each molecule on the basis of its characteristic pattern of peaks, or spectrum. Because the characteristic pattern of all these substances is known, each individual substance can in principle be identified in the overall spectrum of the area of the brain that has been measured. An important metabolite whose concentration is reduced in MCI is N-acetylaspartate (NAA). NAA can be found in healthy, functioning neurons. A reduction in its concentration is seen as evidence for damage to these cells.

3 In 2010, researchers led by Dr Ellen Nollen at UMCG discovered a gene that promotes the aggregation of misfolded proteins. If this gene is deactivated, the aggregation of proteins declines by 75 per cent, which offers scope for treatment. The gene in question is known as MOAG4. Dr Nollen studied this process in tiny worms (*C. elegans*). The leap from worms to humans and a possible treatment for Alzheimer's disease seems huge, and it always takes time before a discovery of this kind leads to effective treatment. But because the ageing process is very fast (only a few weeks) in *C. elegans*, this little nematode offers great opportunities to study the genes and proteins in question.

4 The PET — or positron emission tomography — scan involves the injection (usually into the blood) of a radioactive isotope. As this decays, it sends signals that can be detected by the scanner. These signals are visible on specific receptors in the brain to which the

isotope attaches itself. Receptors are molecules that receive chemical signals from neurons. A specific isotope called ^{11}C-PiB is sensitive to accumulations of beta-amyloid, the protein that becomes misfolded and forms plaques in Alzheimer's disease.

5 The values of t-tau, Aß1–42, and p-tau181p; and the ratios t-tau/Aß1–42 and p-tau181/pAß1–42.

6 That there is a relationship between factors such as proteins in cerebrospinal fluid and performance in neuropsychological tests was also demonstrated in an interesting Swedish study. The researchers compared 73 people with MCI who had normal protein concentrations in their cerebrospinal fluid with 73 who also had MCI but had abnormal protein concentrations. There was a control group consisting of 50 healthy peers. The three groups were exhaustively tested on speed of processing, attention, memory, spatial insight, language, and executive functions. The MCI patients with abnormal protein concentrations performed significantly worse in the tests than the MCI patients with normal protein concentrations, especially on speed of processing and memory. These findings support the idea that abnormal protein concentrations in cerebrospinal fluid and poorer performance on neuropsychological tests reflect the disease process in the brain.

7 The two sorts of medication are acetylcholinesterase inhibitors, which promote the neurotransmission of acetylcholine, and an NMDA receptor antagonist, which blocks the effect of glutamate on brain cells.

8 A. Wiesman, interview with Rudi Westendorp in *de Volkskrant*, 22 October 2011.

9 'Ronald Reagan's son: Alzheimer's seen during presidency', *Reuters*, 14 January 2011.

10 Olde Rikkert M., M. Verbeek, F. Verhey, M. de Vugt, 'De ziekte van Alzheimer bestaat niet' ('Alzheimer's disease doesn't exist'), *NRC Handelsblad*, 11 April 2012.

11 There are indications that Souvenaid (produced by Nutricia), a supplement with a combination of vitamins and amino acids thought to support brain function, may improve memory in Alzheimer's patients. Independent research is needed to confirm this.

CHAPTER 5

1 Simply prescribing a medication for someone can lead to an improvement in their symptoms. Though the improvement may be unconnected with the effectiveness of the drug, it may be wrongly

ascribed to the drug. The placebo effect (an improvement when a treatment containing no active medication has been given) is the result of the patient's positive expectations, mostly unconscious. The subjects in this study did not of course know whether the powder they took every day was the soy preparation or a placebo. Nor did the lead investigators know. This is known as a double-blind study: neither the subjects nor the researchers know which group any subject belongs to. This is vital, because researchers can influence subjects without being aware of it if they have certain expectations.

2 This was a joint study conducted by Dr Marielle Emmelot-Vonk, Dr Harald Verhaar (both colleagues from the geriatrics department of UMCU), and myself, and led by Professor Yvonne van der Schouw.

3 We consulted Edward de Haan, professor of neuropsychology, on the choice of neuropsychological tests. He gave us a very valuable piece of scientific advice that had not occurred to me. I thought that only tests measuring the mental abilities that decline with age were relevant, as you can see whether the results are linked to IGF levels. But Professor de Haan pointed out that if you think that the link between cognitive function and IGF is associated with age-related decline in both, you are not paying attention to the abilities that do *not* decline as we age. As we saw in chapter 1, some things (such as vocabulary, text comprehension, and general knowledge) do not deteriorate as we grow older. So you have a stronger scientific test of your hypothesis if in addition to tests of memory, attention, and cognitive flexibility (what you might call the 'perishables'), you also test vocabulary, reading, and general knowledge (the 'non-perishables'). You would expect a link between IGF levels and perishables, but not between IGF levels and non-perishables. In the end, I administered eight tests: four of abilities that are vulnerable to ageing and four of abilities that are not.

CHAPTER 6

1 Ingmar Vriesema, 'Je moet je blijven ontplooien' ('You have to keep developing your skills'), *NRC Handelsblad*, 18 April 2011.

2 Vitamin B12 (also known as cobalamine) plays a key role in the production of red blood cells and the normal functioning of the nervous system. The only sources of the vitamin are animal products (milk, milk products, meat, meat products, fish, and eggs) or cereals fortified with B12. People at greater risk of B12 deficiency include vegans (as they eat no animal products), older people with a disease

called atrophic gastritis, and those suffering from a deficiency of the stomach protein known as intrinsic factor. A shortage of B12 leads to pernicious anaemia, and may result in neurological symptoms such as paraesthesia (tingling) in the extremities, forgetfulness, ataxia (co-ordination problems), and muscle weakness in the legs.

3 The active ingredients in ginkgo extract are flavonoids and the terpenes ginkgolide and bilobalide.

4 In an article in *Het Parool* newspaper.

CHAPTER 7

1 This is a variation on a proposal for a definition made by Caroline Bassett of The Wisdom Institute (see www.secondjourney.org/itin/09_Fall/Bassett_09Fall.htm).

2 Proverbs 8:1, 10–11 (King James Version).

3 Job 12:12 (King James Version).

4 Powell (2011).

5 Sinnott (1998).

6 The example of Harold and Hanneke is taken, with a few modifications, from Erber (2010).

CHAPTER 8

1 Quoted in Julia Burton-Jones, *Now and Forever: reflections on the later years of life*, Triangle, London, 1997.

2 www.theguardian.com/news/2000/may/19/guardianobituaries.religion

3 This may seem to contradict the research results outlined in chapter 2 showing that older people responded with less brain activity to positive stimulation. But there are two important differences between the study cited in chapter 2 and this study. First, the study in chapter 2 relates to the outside of the prefrontal cortex; this study, to the inside (medial prefrontal cortex and the anterior cingulate). Second, the present study involved happy faces appearing on a screen during performance of a working-memory task. The faces were in the background, so subjects were not obliged to look at them. The fact that emotionally stable older people responded to this background stimulation with more activity in the area of the brain that regulates attention may mean that they pay more attention to positive stimuli.

BIBLIOGRAPHY

GENERAL

Craik, F.I.M., T.A. Salthouse (eds), *The Handbook of Aging and Cognition*, 3rd edition, Psychology Press, New York, 2010, 1–54.

Depp, C.A., D.V. Jeste (eds), *Successful Cognitive and Emotional Aging*, American Psychiatric Publishing, Washington, 2010.

Salthouse, T.A., *Major Issues in Cognitive Aging*, Oxford University Press, Oxford, 2010.

INTRODUCTION

Column 'Het Laatste Woord' (The Last Word), *NRC Handelsblad*, 16–17 June 2012.

CHAPTER 1

Bunce, D., A. Macready, 'Processing Speed, Executive Function, and Age Differences in Remembering and Knowing', *Quarterly Journal of Experimental Psychology*, 2005, 58: 155–168.

Cragg, L., K. Nation, 'Self-Ordered Pointing as a Test of Working Memory in Typically Developing Children', *Memory*, 2007, 15(5): 526–535.

Helmuth, L., 'Aging. The Wisdom of the Wizened', *Science*, 2003, 299: 1300–1302.

Holahan, C.K., C.J. Holahan, K.E. Velasquez, R.J. North, 'Longitudinal Change in Happiness during Aging: the predictive role of positive expectancies', *International Journal of Aging and Human Development*, 2008, 66(3): 229–241.

Hsu, L.M., J. Chung, E.J. Langer, 'The Influence of Age-Related Cues on Health and Longevity', *Perspectives on Psychological Science*, 2010, 5: 632–648.

Kotter-Grühn, D., A. Kleinspehn-Ammerlahn, D. Gerstorf, J. Smith, 'Self-Perceptions of Aging Predict Mortality and Change with Approaching Death: 16-year longitudinal results from the Berlin Aging Study', *Psychology and Aging*, 2009, 24: 654–667.

Levy, B., E. Lange, 'Aging Free from Negative Stereotypes: successful memory in China and among the American deaf', *Journal of*

Personality and Social Psychology, 1994, 66(6): 989–997.

Levy, B.R., M.D. Slade, S.R. Kunkel, S.V. Kasl, 'Longevity Increased by Positive Self-Perceptions of Aging', *Journal of Personality and Social Psychology*, 2002, 83: 261–270.

Magalhães, S.S., A.C. Hamdan, 'The Rey Auditory Verbal Learning Test: normative data for the Brazilian population and analysis of the influence of demographic variables', *Psychology & Neuroscience*, 2010, 3(1): 85–91.

McDaniel, M.A., G.O. Einstein, L.L. Jacoby, 'New Considerations in Aging and Memory', in Craik, F.I.M, T.A. Salthouse (eds), *The Handbook of Aging and Cognition*, Psychology Press, New York, 2008, 251–310.

Salthouse, T.A., *Major Issues in Cognitive Aging*, Oxford University Press, New York, 2010.

Salthouse, T.A., 'Does the Level At Which Cognitive Change Occurs Change with Age?', *Psychological Science*, 2012, 23(1): 18–23.

Salthouse, T.A., J.E. Pink, E.M. Tucker-Drob, 'Contextual Analysis of Fluid Intelligence', *Intelligence*, 2008, 36(5): 464–486.

Society for Experimental Biology, 'Web Weaving Skills Provide Clues to Aging, Spider Study Reveals', *ScienceDaily*, 2 July 2011, www.sciencedaily.com/releases/2011/07/110701203728.htm.

Spreen, O., E. Strauss, *A Compendium of Neuropsychological Tests: administration, norms, and commentary*, 2nd edition, Oxford University Press, New York, 1998.

Thomése, F., A. Bergsma, 'Van Oude Mensen en Dingen Die Nog Komen' (On Loneliness in Older People), *Gerontologie*, 2008, 10(3): 60–63.

Tombaugh, T.N., P. Faulkner, A.M. Hubley, 'Effects of Age on the Rey-Osterrieth and Taylor Complex Figures: test-retest data using an intentional learning paradigm', *Journal of Clinical and Experimental Neuropsychology*, 1992, 14(5): 647–61.

Xu, J., R.E. Roberts, 'The Power of Positive Emotions: it's a matter of life or death — subjective well-being and longevity over 28 years in a general population', *Health Psychology*, 2010, 29(1): 9–19.

CHAPTER 2

Almeida, D.M., M.C. Horn, 'Is Daily Life More Stressful during Middle Adulthood?', in Brim, O.G., C.D. Ryff, R.C. Kessler (eds), *How Healthy Are We? A National Study of Well-Being at Midlife*, University of Chicago Press, Chicago, 2004, 425–451.

Anecdote related by Henk Spaan to Arno Kantelberg in *de Volkskrant*, 28 December 2011.

Ayuso-Mateos, J.L., J.L. Vázquez-Barquero, C. Dowrick, et al., 'Depressive Disorders in Europe: prevalence figures from the ODIN Study', *British Journal of Psychiatry*, 2001, 179: 308–316.

Beekman, A.T.F., 'Neuropathological Correlates of Late-Life Depression', *Expert Reviews in Neurotherapeutics*, 2011, 11: 947–949.

Berk, L.E., *Development through the Lifespan*, 4th edition, Pearson/Allyn & Bacon, Boston, 2007.

Brassen, S., M. Gamer, C. Büchel, 'Anterior Cingulate Activation Is Related to a Positivity Bias and Emotional Stability in Successful Aging', *Biological Psychiatry*, 2011, 70(2): 131–137.

Charles, S.T., L.J. Carstensen, 'Social and Emotional Aging', *Annual Review of Psychology*, 2009, 61: 383–409.

Charles, S.T., B.N. Horwitz, 'Positive Emotions and Health', in Depp, C.A., D.V. Jeste (eds), *Successful Cognitive and Emotional Aging*, American Psychiatric Publishing, Washington, 2010.

Erber, J.T., *Aging and Older Adulthood*, Wiley-Blackwell, Oxford, 2010.

Gross, J.J., L.L. Carstensen, M. Pasupathi, J. Tsai, C.G. Skorpen, A.Y.C. Hsu, 'Emotion and Aging: experience, expression, and control', *Psychology and Aging*, 1997, 12(4): 590–599.

Grossmann, I., J. Na, M.E. Varnum, D.C. Park, S. Kitayama, R.E. Nisbett, 'Reasoning about Social Conflicts Improves into Old Age', *Proceedings of the National Academy of Sciences of the USA*, 2010, 107(16): 7246–7250.

Nashiro, K., M. Sakaki, M. Mather, 'Age Differences in Brain Activity during Emotion Processing: reflections of age-related decline or increased emotion regulation?', *Gerontology*, 2012, 58(2): 156–163.

Scheibe, S., L.L. Carstensen, 'Emotional Aging: recent findings and future trends', *Journals of Gerontology, Series B*, 2010, 65B(2), 135–144.

Sharot, T., A.M. Riccardi, C.M. Raio, E.A. Phelps, 'Neural Mechanisms Mediating Optimism Bias', *Nature*, 2007, 450(7166): 102–105.

Sturrock, D., *Storyteller: the authorized biography of Roald Dahl*, Simon & Schuster, New York, 2010.

Urry, H.L., J.J. Gross, 'Emotion Regulation in Older Age. Current Directions', *Psychological Science*, 2010, 19: 352–357.

Williams, L.M., K.J. Brown, D. Palmer, B.J. Liddell, A.H. Kemp, G. Olivieri, A. Peduto, E. Gordon, 'The Mellow Years? Neural Basis of Improving Emotional Stability over Age', *Journal of Neuroscience*, 2006, 26(24): 6422–6430.

Burgmans, S., E.H. Gronenschild, Y. Fandakova, Y.L. Shing, M.P. van Boxtel, E.F. Vuurman, H.B. Uylings, J. Jolles, N. Raz, 'Age Differences in Speed of Processing Are Partially Mediated by Differences in Axonal Integrity', *NeuroImage*, 2011, 55(3): 1287–1297.

Cabeza, R., S.M. Daselaar, F. Dolcos, S.E. Prince, M. Budde, L. Nyberg, 'Task-Independent and Task-Specific Age Effects on Brain Activity during Working Memory, Visual Attention and Episodic Retrieval', *Cerebral Cortex*, 2004, 14: 364–375.

Christensen, H., K.J. Anstey, L.S. Leach, A.J. Mackinnon, 'Intelligence, Education, and the Brain Reserve Hypothesis', in Craik, F.I.M., T.A. Salthouse (eds), *The Handbook of Aging and Cognition*, 3rd edition, Psychology Press, New York, 2010, 133–188.

Davis, S.W., N.A. Dennis, S.M. Daselaar, M.S. Fleck, R. Cabeza, 'Qué Pasa? The Posterior–Anterior Shift in Aging', *Cerebral Cortex*, 2008, 18(5): 1201–1209.

Dennis, N.A., R. Cabeza, 'Neuroimaging of Healthy Cognitive Aging', 2008, in Craik, F.I.M., T.A. Salthouse (eds), *The Handbook of Aging and Cognition*, 3rd edition, Psychology Press, New York, 2010, 1–54.

Dumas, J.A., P.A. Newhouse, 'The Cholinergic Hypothesis of Cognitive Aging Revisited Again: cholinergic functional compensation', *Pharmacology Biochemistry & Behavior*, 2011, 99(2): 254–261.

Fotenos, A.F., M.A. Mintun, A.Z. Snyder, J.C. Morris, R.L. Buckner, 'Brain Volume Decline in Aging: evidence for a relation between socioeconomic status, preclinical Alzheimer disease, and reserve', *Archives of Neurology & Psychiatry*, 2008, 65(1): 113–120.

Fusco, S., C. Ripoli, M.V. Podda, et al., 'A Role for Neuronal cAMP Responsive-Element Binding (CREB)-1 in Brain Responses to Calorie Restriction', *Proceedings of the National Academy of Sciences of the USA*, 2012, 109(2): 621–626.

Galvan, V., K. Jin, 'Neurogenesis in the Aging Brain', *Journal of Clinical Interventions in Aging*, 2007, 2 (4): 605–610.

Guttman, M., 'The Aging Brain', *USC Health Magazine*, 2001, Spring.

Penke, L., S. Muñoz Maniega, C. Murray, A.J. Gow, M.C. Valdèz Hernández, J.D. Clayden, J.M. Starr, J.M. Wardlaw, M.E. Bastin, I.J. Deary, 'A General Factor of Brain White Matter Integrity Predicts Information Processing Speed in Healthy Older People', *Journal of Neuroscience*, 2010, 30(22): 7569–7574.

Resnick, S.M., D.L. Pham, M.A. Kraut, A.B. Zonderman, C. Davatzikos, 'Longitudinal Magnetic Resonance Imaging Studies of Older Adults: a shrinking brain', *Journal of Neuroscience*, 2003, 23(8): 3295–3301.

Resnick, S.M., J. Sojkova, Y. Zhou, Y. An, W. Ye, D.P. Holt, R.F. Dannals, C.A. Mathis, W.E. Klunk, L. Ferrucci, M.A. Kraut, D.F. Wong, 'Longitudinal Cognitive Decline Is Associated with Fibrillar Amyloid-Beta Measured by [11C]PiB', *Neurology*, 2010, 74(10): 807–815.

Salthouse, T.A., 'Neuroanatomical Substrates of Age-Related Cognitive Decline', *Psychological Bulletin*, 2011, 137: 753–784.

Shors, T.J., G. Miesegaes, A. Beylin, M. Zhao, T. Rydel, E. Gould, 'Neurogenesis in the Adult Is Involved in the Formation of Trace Memories', *Nature*, 2001, 410 (6826): 372–376. Erratum in *Nature*, 414(6866): 938.

Sowell, E.R, P.M. Thompson, A.W. Toga, 'Mapping Changes in the Human Cortex throughout the Span of Life', *Neuroscientist*, 2004, 10: 372–392.

Villeda, S.A., J. Luo, K.I. Mosher, et al., 'The Ageing Systemic Milieu Negatively Regulates Neurogenesis and Cognitive Function', *Nature*, 2011, 477(7362): 90–94.

CHAPTER 4

Advice on MCI: Mayoclinic.

Baker, L.D., L.L. Frank, K. Foster-Schubert, et al., 'Effects of Aerobic Exercise on Mild Cognitive Impairment: a controlled trial', *Archives of Neurology and Psychiatry*, 2010, 67(1): 71–79.

Belleville, S. 'Cognitive Training for Persons with Mild Cognitive Impairment', *International Psychogeriatrics*, 2008, 20: 57–66.

Binnewijzend, M.A., M.M. Schoonheim, E. Sanz-Arigita, A.M. Wink, W.M. van der Flier, N. Tolboom, S.M. Adriaanse, J.S. Damoiseaux, P. Scheltens, B.N. van Berckel, F. Barkhof, 'Resting-State fMRI Changes in Alzheimer's Disease and Mild Cognitive Impairment', *Neurobiology of Aging*, Sept. 2012, 33(9): 2018–2028.

Buschert, V., A.L. Bokde, H. Hampel, 'Cognitive Intervention in Alzheimer Disease', *Nature Reviews Neurology*, 2010, 6(9): 508–517.

Carlson, M.C., K.I. Erickson, A.F. Kramer, et al., 'Evidence for Neurocognitive Plasticity in At-Risk Older Adults: the Experience Corps Program', *Journals of Gerontology, Series A: Biological Sciences and Medical Sciences*, 2009, 64(12): 1275–1282.

Cherbuin, N., P. Sachdev, K.J. Anstey, 'Neuropsychological Predictors of Transition from Healthy Cognitive Aging to Mild Cognitive

Impairment: the PATH through life study', *American Journal of Geriatric Psychiatry*, 2010, 18(8): 723–733.

Costafreda, S.G., I.D. Dinov, Z. Tu, et al., 'Automated Hippocampal Shape Analysis Predicts the Onset of Dementia in Mild Cognitive Impairment', *NeuroImage*, 2011, 56(1): 212– 219.

Cui, Y., B. Liu, S. Luo, et al., 'Identification of Conversion from Mild Cognitive Impairment to Alzheimer's Disease Using Multivariate Predictors', *PLoS One*, 2011, 6(7): e21896 [ePub].

Erk, S., A. Spottke, A. Meisen, et al., 'Evidence of Neuronal Compensation during Episodic Memory in Subjective Memory Impairment', *Archives of General Psychiatry*, 2011, 68(8): 845–852.

Ferreira, L.K., B.S. Diniz, O.V. Forlenza, et al., 'Neurostructural Predictors of Alzheimer's dDisease: a meta-analysis of VBM studies', *Neurobiology of Aging*, 2011, 32(10): 1733–1741.

Gates, N.J., P.S. Sachdev, M.A. Fiatarone Singh, M. Valenzuela, 'Cognitive and Memory Training in Adults at Risk of Dementia: a systematic review', *BMC Geriatrics*, 2011, 11: 55.

Geda, Y.E., H.M. Topazian, R.A. Lewis, et al., 'Engaging in Cognitive Activities, Aging, and Mild Cognitive Impairment: a population-based study', *Journal of Neuropsychiatry and Clinical Neurosciences*, 2011, 23(2): 149–54.

Li, H., J. Li, N. Li, B. Li, et al., 'Cognitive Intervention for Persons with Mild Cognitive Impairment: a meta-analysis', *Ageing Research Reviews*, 2011, 10(2): 285–296.

Olde Rikkert, M., M. Verbeek, F. Verhey, M. De Vugt, 'De Ziekte van Alzheimer Bestaat Niet' (Alzheimer's Disease Does Not Exist), *NRC Handelsblad*, 11 April 2012.

Powell, H.D., 'Not Everyone with Mild Cognitive Impairment Progresses to Dementia', *Psychology Today*, 2 June 2011, www. psychologytoday.com/blog/the-aging-intellect/201106/not-everyone-mild-cognitive-impairment-progresses-dementia.

Ronald Reagan's public letter about his Alzheimer's: www.reagan. utexas.edu/archives/reference/alzheimerletter.html.

Yankner, B.A., T. Lu, P. Loerch, 'The Aging Brain', *Annual Review of Pathology*, 2008, 3: 41–66.

CHAPTER 5

Aleman, A., I. Torres-Alemán, 'Circulating Insulin-Like Growth Factor I and Cognitive Function: neuromodulation throughout the lifespan', *Progress in Neurobiology*, 2009, 89(3): 256–265.

Aleman, A., H.J. Verhaar, E.H. de Haan, W.R. de Vries, M.M. Samson, M.L. Drent, E.A. van der Veen, H.P. Koppeschaar, 'Insulin-Like Growth Factor-1 and Cognitive Function in Healthy Older Men', *Journal of Clinical Endocrinology & Metabolism*, 1999, 84(2) 471–475.

Ayers, B., M. Forshaw, M.S. Hunter, 'The impact of Attitudes towards the Menopause on Women's Symptom Experience: a systematic review', *Maturitas*, 2010, 65(1): 28–36.

Dik, M.G., S.M. Pluijm, C. Jonker, D.J. Deeg, M.Z. Lomecky, P. Lips, 'Insulin-Like Growth Factor I (IGF-I) and Cognitive Decline in Older Persons', *Neurobiology of Aging*, 2003, 24(4): 573–81. Erratum in *Neurobiology of Aging*, Feb. 2004, 25(2): 271.

Dumas, J.A., A.M. Kutz, M.R. Naylor, J.V. Johnson, P.A. Newhouse, 'Increased Memory Load-Related Frontal Activation after Estradiol Treatment in Postmenopausal Women', *Hormones and Behavior*, 2010, 58(5): 929–935.

Emmelot-Vonk, M.H., H.J. Verhaar, H.R. Nakhai Pour, A. Aleman, T.M. Lock, J.L. Bosch, D.E. Grobbee, Y.T. van der Schouw, 'Effect of Testosterone Supplementation on Functional Mobility, Cognition, and Other Parameters in Older Men: a randomized controlled trial', *Journal of the American Medical Association*, 2008, 299(1): 39–52.

Foster, T.C., 'Role of Estrogen Receptor Alpha and Beta Expression and Signaling on Cognitive Function during Aging', *Hippocampus*, 2012, 22(4): 656–669.

Hogervorst, E., S. Bandelow, 'Sex Steroids to Maintain Cognitive Function in Women after the Menopause: a meta-analysis of treatment trials', *Maturitas*, 2010, 66(1): 56–71.

Holland, J., S Bandelow, E. Hogervorst, 'Testosterone Levels and Cognition in Elderly Men: a review', *Maturitas*, 2011, 69(4): 322–337.

Kreijkamp-Kaspers, S., L. Kok, D.E. Grobbee, E.H. de Haan, A. Aleman, Y.T. van der Schouw, 'Dietary Phytoestrogen Intake and Cognitive Function in Older women', *Journals of Gerontology, Series A: Biological Sciences and Medical Sciences*, 2007, 62(5): 556–562.

Kreijkamp-Kaspers, S., L. Kok, D.E. Grobbee, E.H. de Haan, A. Aleman, J.W. Lampe, Y.T. van der Schouw, 'Effects of Soy Protein Containing Isoflavones on Cognitive Function, Bone Mineral Density and Plasma Lipids in Postmenopausal Women: a randomized trial', *Journal of the American Medical Association*, 2004, 292(1): 65–74.

Leon-Carrion, J., J.F. Martin-Rodriguez, A. Madrazo-Atutxa, A. Soto-Moreno, E. Venegas-Moreno, E. Torres-Vela, P. Benito-López, M.A. Gálvez, F.J. Tinahones, A. Leal-Cerro, 'Evidence of Cognitive and Neurophysiological Impairment in Patients with Untreated Naive Acromegaly', *Journal of Clinical Endocrinology & Metabolism*, 2010, 95(9): 4367–4379.

Melby, M.K., M. Lock, P. Kaufert, 'Culture and Symptom Reporting at Menopause', *Human Reproduction Update*, 2005, 11(5): 495–512.

Muller, M, A. Aleman, D.E. Grobbee, E.H. de Haan, Y.T. van der Schouw, 'Endogenous Sex Hormone Levels and Cognitive Function in Aging Men: is there an optimal level?', *Neurology*, 2005, 64(5): 866–871.

Shepard, R. N., J. Metzler, 'Mental Rotation of Three-Dimensional Objects', *Science*, 2005, 171: 701–703.

CHAPTER 6

Aleman, A., *Natuurlijk Beter; Natuurlijke Middelen voor Neerslachtigheid, Nervositeit, Vergeetachtigheid en Slapeloosheid* (about the use of natural supplements to combat anxiety and insomnia), Den Hertog, Houten, 2004.

Angevaren, M., G. Aufdemkampe, H.J. Verhaar, A. Aleman, L. Vanhees, 'Physical Activity and Enhanced Fitness to Improve Cognitive Function in Older People without Known Cognitive Impairment', *Cochrane Database of Systematic Reviews*, 16 April 2008, (2): cd005381. 43.

Baker, K., 'Musical Activity May Help the Aging Brain', *Emory Report*, 22 April 2011, www.emory.edu/EMORY_REPORT/stories/2011/04/research_musical_activity_aging_brain.html.

Basak, C., W.R. Boot, M.W. Voss, A.F. Kramer, 'Can Training in a Real-Time Strategy Video Game Attenuate Cognitive Decline in Older Adults?', *Psychology and Aging*, 2008, 23: 765–777.

Cicero, Marcus Tullius, *De Senectute* (On Old Age), translated by Andrew P. Peabody, Little, Brown, and Co., Boston, 1887.

Colcombe, S.J., K.I. Erickson, P.E. Scalf, J.S. Kim, R. Prakash, E. McAuley, S. Elavsky, D.X. Marquez, L. Hu, A.F. Kramer, 'Aerobic Exercise Training Increases Brain Volume in Aging Humans', *Journals of Gerontology, Series A: Biological Sciences and Medical Sciences*, 2006, 61(11): 1166–1170.

Dohmen J., J. Baars (eds), *De Kunst van het Ouder Worden; de Grote Filosofen over Ouderdom* (Views of the Great Philosophers on Ageing), 2nd edition, Ambo, Amsterdam, 2010.

Durga, J., M.P. van Boxtel, E.G. Schouten, F.J. Kok, J. Jolles, M.B. Katan, P. Verhoef, 'Effect of 3-Year Folic Acid Supplementation on Cognitive Function in Older Adults in the FACIT Trial: a randomised, double blind, controlled trial', *Lancet*, 2007, 369(9557), 208–216.

Erickson, K.I., M.W. Voss, R.S. Prakash, C. Basak, A. Szabo, L. Chaddock, J.S. Kim, S. Heo, H. Alves, S.M. White, T.R. Wojcicki, E. Mailey, V.J. Vieira, S.A Martin, B.D. Pence, J.A. Woods, E. McAuley, A.F. Kramer, 'Exercise Training Increases Size of Hippocampus and Improves Memory', *Proceedings of the National Academy of Sciences of the USA*, 2011, 108(7): 3017–3022.

Fernández-Prado, S. Conlon, J.M. Mayán-Santos, M. Gandoy-Crego, 'The Influence of a Cognitive Stimulation Program on the Quality of Life Perception among the Elderly', *Archives of Gerontology and Geriatrics*, 2012, 54(1): 181–184.

FitzGerald, D.B, G.P. Crucian, J.B. Mielke, B.V. Shenal, D. Burks, K.B. Womack, G. Ghacibeh, V. Drago, P.S. Foster, E. Valenstein, K.M. Heilman, 'Effects of Donepezil on Verbal Memory after Semantic Processing in Healthy Older Adults', *Cognitive and Behavioral Neurology*, 2008, 21(2): 57–64.

Gao, Q., M. Niti, L. Feng, K.B. Yap, T.P. Ng, 'Omega-3 Polyunsaturated Fatty Acid Supplements and Cognitive Decline: Singapore Longitudinal Aging Studies', *Journal of Nutrition, Health and Aging*, 2011, 15(1): 32–35.

Gross, A.L., G.W. Rebok, 'Memory Training and Strategy Use in Older Adults: results from the active study', *Psychology and Aging*, 2011, 26(3): 503–517.

Hindin, S.B., E.M. Zelinski, 'Extended Practice and Aerobic Exercise Interventions Benefit Untrained cognitive outcomes in older adults: a meta-analysis', *Journal of the American Geriatrics Society*, Jan. 2012, 60(1): 136–141.

Hornung, O.P., F. Regen, H. Danker-Hopfe, M. Schredl, I. Heuser, 'The Relationship between REM Sleep and Memory Consolidation in Old Age and Effects of Cholinergic Medication', *Biological Psychiatry*, 2007, 61(6): 750–757.

Lövdén, M., N.C. Bodammer, S. Kühn, J. Kaufmann, H. Schütze, C. Tempelmann, H.J. Heinze, E. Düzel, F. Schmiedek, U.

Lindenberger, 'Experience-Dependent Plasticity of White-Matter Microstructure Extends into Old Age', *Neuropsychologia*, 2010, 48(13): 3878–3883.

Lustig, C., P. Shah, R. Seidler, P.A. Reuter-Lorenz, 'Aging, Training, and the Brain: a review and future directions', *Neuropsychology Review*, 2009, 19(4): 504–522.

Mozaffarian, D., J.H. Wu, 'Omega-3 Fatty Acids and Cardiovascular Disease: effects on risk factors, molecular pathways, and clinical events', *Journal of the American College of Cardiology*, 2011, 58(20): 2047–2067.

Mozolic, J.L., S. Hayasaka, P.J. Laurienti, 'A Cognitive Training Intervention Increases Resting Cerebral Blood Flow in Healthy Older Adults', *Frontiers in Human Neuroscience*, 2010, 4: 16.

Parker, S., M. Garry, G.O. Einstein, M.A. McDaniel, 'A Sham Drug Improves a Demanding Prospective Memory Task', *Memory*, 2011, 19(6): 606–612.

Rapoport, M.J., B. Weaver, A. Kiss, C. Zucchero Sarracini, H. Moller, N. Herrmann, K. Lanctôt, B. Murray, M. Bédard, 'The Effects of Donepezil on Computer-Simulated Driving Ability among Healthy Older Adults: a pilot study', *Journal of Clinical Psychopharmacology*, 2011, 31(5): 587–592.

Silberstein, R.B., A. Pipingas, J. Song, D.A. Camfield, P.J. Nathan, C. Stough, 'Examining Brain-Cognition Effects of Ginkgo Biloba Extract: brain activation in the left temporal and left prefrontal cortex in an object working memory task', *Evidence-Based Complementary and Alternative Medicine*, 2011: 164139, 18 August 2011 [ePub].

Singh-Manoux, A., M. Kivimaki, M.M. Glymour, A. Elbaz, C. Berr, K.P. Ebmeier, J.E. Ferrie, A. Dugravot, 'Timing of Onset of Cognitive Decline: results from Whitehall II prospective cohort study', *British Medical Journal*, 2011, 344: d7622.

Smith, A.D., H. Refsum, 'Vitamin B-12 and Cognition in the Elderly', *American Journal of Clinical Nutrition*, 2009, 89(2): 707S–711S.

Turner, D.C., L. Clark, J. Dowson, T.W. Robbins, B.J. Sahakian, 'Modafinil improves cognition and response inhibition in adult attention-deficit/hyperactivity disorder', *Biological Psychiatry*, 15 May 2004, 55(10): 1031–1040.

Turner, D.C., L. Clark, E. Pomarol-Clotet, P. McKenna, T.W. Robbins, B.J. Sahakian, 'Modafinil Improves Cognition and Attentional Set Shifting in Patients with Chronic Schizophrenia',

Neuropsychopharmacology, 2004, 29(7): 1363–1373.

Turner, D.C., T.W. Robbins, L. Clark, A.R. Aron, J. Dowson, B.J. Sahakian, 'Cognitive Enhancing Effects of Modafinil in Healthy Volunteers', *Psychopharmacology*, 2003, 165(3): 260–269.

Wan, C.Y., G. Schlaug, 'Music Making as a Tool for Promoting Brain Plasticity across the Life Span', *Neuroscientist*, 2010, 16(5): 566–577.

Wisdom of Sirach, chapter 38, verse 4, Revised Standard Version of the Bible.

CHAPTER 7

'Kofi Annan takes over Kenya mediation', *CBS News*, 10 January 2008, www.cbsnews.com/news/kofi-annan-takes-over-kenya-mediation-10-01-2008/.

Ardelt, M., M.S.W. Hunhui Oh, 'Wisdom', in Depp, C.A., D.V. Jeste (eds), *Successful Cognitive and Emotional Aging*, American Psychiatric Publishing, Washington, 2010.

Cato, M.A., D.C. Delis, T.J. Abildskov, E. Bigler, 'Assessing the Elusive Cognitive Deficits Associated with Ventromedial Prefrontal Damage: a case of a modern-day Phineas Gage', *Journal of the International Neuropsychological Society*, 2004, 10: 453–465.

Cohen, G.D., *The Mature Mind: the positive power of the aging brain*, Basic Books, New York, 2005.

Collins, N., 'Wisdom Comes with Age, Study Shows', *The Telegraph*, 24 August 2011.

Cozolino, L.J., *The Healthy Aging Brain: sustaining attachment, attaining wisdom*, Norton & Company, New York, 2008.

Eppinger, B., J. Kray, B. Mock, A. Mecklinger, 'Better or Worse than Expected? Aging, Learning, and the ERN', *Neuropsychologia*, 2008, 46(2): 521–539.

Glück, J., S. Bluck, 'Laypeople's Conception of Wisdom and Its Development: cognitive and integrative views', *Journals of Gerontology, Series B: Psychological Sciences and Social Sciences*, 2011, 66(3): 321–324.

Goldberg, E., *De Wijsheidparadox; Hoe het Verstand Groeit Terwijl de Hersenen Ouder Worden* (The Wisdom Paradox), Wereldbibliotheek, Amsterdam, 2007.

Helmuth, L., 'Aging. The Wisdom of the Wizened', *Science*, 2003, 299(5611): 1300–1302.

Jeste, D.V., J.C. Harris, 'Wisdom — a Neuroscience Perspective', *Journal of the American Medical Association*, 2010, 304(14): 1602–1603.

Meeks, T.W., D.V. Jeste, 'Neurobiology of Wisdom: a literature overview', *Archives of General Psychiatry*, 2009, 66(4): 355–365.

Powell, D.H., *The Aging Intellect*, Routledge, New York, 2011.

Sinnott, J.D., *The Development of Logic in Adulthood*, Plenum Press, New York, 1998.

Smith, J., P.B. Baltes, 'Wisdom-Related Knowledge: age/cohort differences in response to life-planning problems', *Developmental Psychology*, 1990, 26(3): 494–505.

Staudinger, U.M., J. Glück, 'Psychological wisdom research: commonalities and differences in a growing field', *Annual Review of Psychology*, 2011, 62: 215–241.

Sternberg, R.J., J. Jordan (eds), *A Handbook of Wisdom: psychological perspectives*, Cambridge University Press, Cambridge, 2005.

CHAPTER 8

Blazer, D.G., K.G. Meador, 'The Role of Spirituality in Healthy Aging', in Depp, C.A., D.V. Jeste (eds), *Successful Cognitive and Emotional Aging*, American Psychiatric Publishing, Washington, 2010.

Brassen, S., M. Gamer, C. Büchel, 'Anterior Cingulated Activation Is Related to a Positivity Bias and Emotional Stability in Successful Aging', *Biological Psychiatry*, 2011, 70(2): 131–137.

Depp, C.A., A. Harmell, I.V. Vahia, 'Successful Cognitive Aging', *Current Topics in Behavioral Neuroscience*, 2012, 10: 35–50.

Farrell, M.J., F. Zamarripa, R. Shade, P.A. Phillips, M. McKinley, P.T. Fox, J. Blair-West, D.A. Denton, G.F. Egan, 'Effect of Aging on Regional Cerebral Blood Flow Responses Associated with Osmotic Thirst and Its Satiation by Water Drinking: a PET study', *Proceedings of the National Academy of Sciences of the USA*, 2008, 105(1): 382–387.

Fernandez, A., 'Marian Diamond on the Brain', *Sharp Brains*, 5 November 2007, www.sharpbrains.com/blog/2007/11/05/marian-diamond-on-the-brain/.

Gallucci, M., P. Antuono, F. Ongaro, P.L. Forloni, D. Albani, G.P. Amici, C. Regini, 'Physical Activity, Socialization and Reading in the Elderly over the Age of Seventy: what is the relation with cognitive decline? Evidence from "The Treviso Longeva (TRELONG) Study"', *Archives of Gerontology & Geriatrics*, 2009, 48(3): 284–286.

Geda, Y.E., H.M. Topazian, R.A. Lewis, R.O. Roberts, D.S. Knopman, V.S. Pankratz, T.J. Christianson, B.F. Boeve, E.G. Tangalos, R.J.

Ivnik, R.C. Petersen, 'Engaging in Cognitive Activities, Aging, and Mild Cognitive Impairment: a population-based study', *Journal of Neuropsychiatry & Clinical Neurosciences*, 2011, 23(2): 149–154.

Kahn, R.S., *De Tien Geboden voor het Brein* (Ten Commandments for the Brain), Balans, Amsterdam, 2011.

Kreijkamp-Kaspers, S., L. Kok, D.E. Grobbee, E.H. de Haan, A. Aleman, Y.T. van der Schouw, 'Dietary Phytoestrogen Intake and Cognitive Function in Older Women', *Journals of Gerontology, Series A: Biological Sciences and Medical Sciences*, 2007, 62(5): 556–562.

Lee, B.K., T.A. Glass, B.D. James, K. Bandeen-Roche, B.S. Schwartz, 'Neighborhood Psychosocial Environment, Apolipoprotein E Genotype, and Cognitive Function in Older Adults', *Archives of General Psychiatry*, 2011, 68(3): 314–321.

Li, T., H.H. Fung, D.M. Isaacowitz, 'The Role of Dispositional Reappraisal in the Age-Related Positivity Effect', *Journals of Gerontology, Series B: Psychological Sciences and Social Sciences*, 2011, 66(1): 56–60.

Nussbaum, P., *Save Your Brain: the 5 things you must do to keep your mind young and sharp*, McGraw-Hill, New York, 2010.

Rowe, J.W., R.L. Kahn, *Successful Aging*, Pantheon Books, New York, 1998.

Siegel, D.J., *The Mindful Brain*, Norton, New York, 2007.

Stern, C., Z. Munn, 'Cognitive Leisure Activities and Their Role in Preventing Dementia: a systematic review', *International Journal of Evidence-Based Healthcare*, 2010, 8(1): 2–17.

Willcox, B.J., C.D. Willcox, M. Suzuki, *The Okinawa Way: how to improve your health and longevity dramatically*, Penguin Books, London, 2001.

Wilson, R.S., E. Segawa, P.A. Boyle, D.A. Bennett, 'Influence of Late-Life Cognitive Activity on Cognitive Health', *Neurology*, 2012, 78(15): 1123–1129.

ILLUSTRATION CREDITS

Figure 1: Branislava Curcic-Blake, based on Salthouse et al. (2008).

Figure 2: Based on Larry R. Squire, 'Memory Systems of the Brain: a brief history and current perspective', *Neurobiology of Learning and Memory*, 2004, 82:171-177.

Figure 3: Berber Munstra, for this book. Variation on Rey's complex figure.

Figure 4: From Cragg & Nation (2007).

Figure 5: André Aleman, for this book.

Figure 6: Neurosciences Department, UMCG.

Figure 7: Branislava Curcic-Blake, based on Salthouse et al. (2008).

Figure 8: Branislava Curcic-Blake, based on Salthouse et al. (2008).

Figure 9: Berber Munstra, for this book.

Figure 10: Santiago Ramón y Cajal.

Figure 11: Berber Munstra, for this book.

Figure 12: Berber Munstra, for this book.

Figure 13: Berber Munstra, for this book.

Figure 14: From Cabeza et al. (2004).

Figure 15: Berber Munstra, for this book, based on Yankner et al. (2008).

Figure 16: Berber Munstra, for this book.

Figure 17: From Shepard & Metzler (1971).

Figure 18: André Aleman, for this book.

Figure 19: Berber Munstra, for this book.

Figure 20: Berber Munstra, for this book.

INDEX